ANIMAL PARTNERS
Training Animals To Help People

ANIMAL PARTNERS

Training Animals
To Help People

PATRICIA CURTIS

LODESTAR BOOKS

E. P. DUTTON • NEW YORK

Library of Congress Cataloging in Publication Data

Curtis, Patricia.
    Animal partners.

    "Lodestar books."
    Bibliography: p.
    Includes index.
    Summary: Discusses ways in which animals are trained to help the handicapped and to provide companionship for mentally retarded, ill, and elderly persons. Based on interviews with professional animal trainers, therapists, social workers, and teachers.
    1. Domestic animals—Therapeutic use—Juvenile literature. 2. Rehabilitation—Juvenile literature. 3. Pets—Therapeutic use—Juvenile literature. 4. Working animals—Juvenile literature. 5. Animals, Training of—Juvenile literature. [1. Working animals. 2. Animals—Training. 3. Animals—Therapeutic use. 4. Animal specialists. 5. Occupations] I. Title.
RM931.A65C87 1982 636.08'86 82-12969
ISBN 0-525-66791-1

Published in the United States by E. P. Dutton, Inc., 2 Park Avenue, New York, N.Y. 10016
Published simultaneously in Canada by Clarke, Irwin & Company Limited, Toronto and Vancouver

EDITOR: VIRGINIA BUCKLEY

Printed in the U.S.A.   First Edition
10 9 8 7 6 5 4 3 2 1

to Anna Marra,
with gratitude and affection

CONTENTS

Acknowledgments   ix

Preface   xi

1. My Partner, My Dog   1

2. Horse Power   15

3. Eyes With Four Legs   34

4. Therapists in the Barn   51

5. Pets and the Healing Process   67

6. Ears With Orange Collars   77

7. "Pet Therapy" for the Elderly   95

   Further Information   115

   Index   125

ACKNOWLEDGMENTS

You've heard of the Underground Railroad? It wasn't an actual railroad; it was a network of helping human beings. During slavery days in America, runaway slaves were often helped to freedom by compassionate people who secretly passed them along northward from one to another, each person or "station" giving the escaped slave food, shelter, and directions to the next member of the railroad. Well, writing a book such as this one, about a subject so new, is of course nothing like running for your life—but still in some ways is a little like a trip on the Underground Railroad. People pass you along. They give you information and guidance, and direct you to the next person you should contact to get your story.

Many men and women helped me in this way when I was writing this book. Because much of the work of using animals to help the handicapped is new, and the fields are relatively small, people doing this work often know one another. The people long ago who were "stations" on the Underground Railroad all had a commitment to the idea of human freedom and a revulsion against human bondage. People who train and use animals to help others seem to share qualities of caring, of strong commitment to people, and of deep affection for animals. I feel privileged to have met so many.

I especially wish to express my warmest gratitude to: Martha Biery and Mildred Krentel of the Melmark Home; Marjorie Kittredge of Windrush Farm; Lida L. McCowan of the Cheff Center for the Handicapped; Leo K. Bustad and Linda Hines of the People-Pet Partnership; Ernest M. Swanton of the Guide Dog Foundation; Donald P. MacMunn of the New England Education Center; Agnes McGrath of International Hearing Dog, Inc.; Samuel B. Ross, Jr., of Green Chimneys Farm Center; Sue Myles of Able Dogs; Starr R. Hayes of Aid Dogs for the Handicapped; Aaron H. Katcher of the Center for the Interaction of Animals and Society; Judith Star of the American Humane Education Society; and Judith Feldman of Children and Animals Together for Seniors.

In addition, I wish to thank John Byfield, Joe Capizzo, Earl W. Capron, Ange Condoret, Samuel A. Corson, Debbie Faul, Cathy Gross, Patricia Gwozdz, Heidi Jadelskyj, David Lee, Alexandra Meaders, Lee Mitchell, Catherine Setterlin, Vera Slover, Keith Wilbur, and Marilyn Williams.

The book would have been difficult to write without the encouragement and assistance of Jean Stewart—thank you, Jean—and of Argus Archives, where research material was made available to me. And finally, my thanks to David Reuther for his good advice and encouragement and to Virginia Buckley for her generous help and support.

Patricia Curtis

PREFACE

When I first started this book, I had to ask myself whether or not the use of animals to help disabled people constituted a form of exploitation of the animals. Is this yet another kind of tyranny over animals? We feel free to help ourselves to animals in this work because we need them, and because they cannot stop us from using them.

If you take the position that all work is drudgery, an unpleasant necessity, then I suppose making animals work is bound to seem unkind, even cruel. However, work is not necessarily demeaning or hateful. Making animals work can be adding a dimension to their lives that not only does not harm them but may even make them happy.

Animals have worked for people for thousands of years. Many have been treated with monstrous cruelty; others have been appreciated and cared for. It's hard to tell for sure how an animal feels about working. Yet, anyone who has watched a sheep dog rounding up its flock will agree that the animal seems to enjoy the job. Sled dogs are raring to go. Farm horses that are well treated and well respected do not appear to mind their work.

Puppies and kittens taken to visit elderly people in nursing homes and hospitals love the attention they get if their visits are not overlong. From what I have learned of guide dogs for the deaf and blind, they do not suffer from their work if they are loved and well provided for—in fact, they seem proud of themselves. The trained horses I have seen carrying handicapped people are not exploited but greatly valued, and appear not only to tolerate their work but to enjoy it.

Helping and caring for other people can make us feel great, and nurturing our pets is certainly rewarding. When I take care of my dog and cats—feed, brush, walk, exercise, play with, and clean up after them, take them to the veterinarian if they need it—some of that is work for me, but well worth it. In return, all they have to do is be themselves. If my dog had to take care of me, if I were blind and she were trained to guide me, would she love me any less? Would she be less happy? I doubt it.

I think the key to using animals to help the disabled lies in training them to do only what is well within their abilities and not pushing them beyond their limits. Punishment must never be used in their training. People must never expect animals to perform in ways that make them tense, anxious, exhausted, or unhappy. The animals must always be well cared for and treated with utmost consideration.

The fortunate people who receive the trained animals to be their partners and aides also must be on their guard. The temptation to exploit those who serve us is strong. A person must never allow a disability to become the excuse for overlooking the needs and happiness of his or her helping companion animal.

"I would sound a note of caution," says John A. Hoyt, President of the Humane Society of the United States. "Man's utilization of animals cannot be taken for granted, but must be tempered with compassion, care, and concern. It is of little value that we serve the good on one hand if we do injustice on the other. In the further bonding of humans and animals in new and beneficial ways, not only can the lives of humans be enriched and made whole, but those animals which we use can be afforded genuine love, companionship, and care."

The fields of professional work in training animals for handicapped people are growing, stimulated by the new discoveries and rising scientific interest in the human/companion animal bond, the love between people and their pets. I think our perceptions of the creatures we share the planet with are changing in ways that will benefit all of us.

Young people today who are interested in working with animals will find growing opportunities far beyond the usual fields of veterinary medicine, dog and cat breeding, zoo keeping, and so forth. In the chapters in this book, I have attempted to give you an overview of the requirements, training, advantages, and drawbacks of several different types of work. The young men and women who tell their stories here are composites of real persons.

In addition to professional opportunities, there is much needed volunteer work. Therapeutic horseback riding, for example, calls for well-trained professionals—but some centers also need volunteers as pony leaders and sidewalkers. You need special training to teach dogs for the blind and deaf; however, SPCAs and other humane organizations in many cities need volunteers to assist in the "pet therapy" programs for the elderly in nursing homes and for children in hospitals. High schoolers who help in volunteer programs may well improve their qualifications for professional training in animal-related careers later on.

There is another plus to working with animals in these ways. Training animals for this work increases appreciation for them and raises their status. Working with disabled people will heighten your respect for them and understanding of their problems. Handicapped people are pressing for their rights to function in society in as many ways as possible, and these rights are long overdue. Trained animals can help them. Young people who are attracted to this work first of all because they enjoy being around animals will soon realize an extra inner happiness that comes from helping and learning from other human beings as well.

1
MY PARTNER, MY DOG

Katie tells how
she uses her experience
as a professional dog trainer
to enrich the lives
of handicapped people.

"Sit down, Alfie! Sit!" commanded the little girl.

Alfie wagged his tail, put his paws on her lap, and gave her a big wet kiss. He was a good-natured, lovable dog, but not easily impressed when it came to commands.

"He won't do it, Katie," said the child to me. She kept trying to pull on Alfie's leash.

I watched, wondering how I was going to solve this one. First of all, Alfie was an enormous Old English sheepdog, and Louise was only seven years old. Secondly, Louise was in a wheelchair, paralyzed from the waist down, and her voice was thin and didn't carry much authority. Nevertheless, she was a member of my dog-training class for disabled children and had brought her dog Alfie to learn obedience. It was important that Louise have the satisfaction of training her dog.

1

Well, Alfie will just have to learn to sit in response to a hand signal, that's all, I thought. In all my classes, I improvise freely, adapting or changing the tried-and-true methods I use as a professional dog trainer, or inventing new methods altogether. I also adjust my expectations according to the abilities of each student and each dog.

There were five students in this class—three young teenagers, a ten-year-old, and Louise. All were in wheelchairs with varying degrees of disability. Their dogs ranged in size and type from Alfie to a miniature poodle, and while I felt that some of them were ideal for obedience training, there were a couple I had my doubts about.

No matter—turning out perfectly obedient dogs was not my objective with this group. One important thing I wanted to accomplish was to have the children teach their dogs enough to give the kids a feeling of power and responsibility.

Unfortunately, our society has traditionally tended to exclude handicapped people from the mainstream of life and to keep them isolated. Though handicapped people themselves are now pressing actively for their rights to have the same opportunities as everyone else, the discrimination they face is still great. They are often excluded from jobs that they could perform perfectly well. They may be denied houses or apartments. And often people who aren't disabled feel embarrassed and uncomfortable around handicapped people.

Well, teaching children to train their dogs will not change all that, but it will give the kids a feeling of accomplishment. The dogs will mean more to them. And if they compete with their dogs in the obedience classes in dog

2

shows, they compete as equals with the other dog owners. That's something.

I had been majoring in education in college, planning to become a teacher, when I realized that what I really wanted to do was to work professionally with dogs. So now I earn my living as a dog trainer.

How did I hit on the idea of combining my experience as a trainer with my interest in teaching, and wind up working with the handicapped? It came about by chance. I have a friend, Patrick, who's a staff photographer on our local newspaper, and sometimes if I'm not busy I go with him on shootings. One Saturday afternoon he called me up.

"I have to go photograph a guy and his dog," said Patrick. "It's quite a story—the dog saved the man's life by dragging him out of a fire. The man is paralyzed, lives alone with his dog. A reporter has done a story on him and his dog and needs a photo to go with it. Want to come along?"

I had no more classes to teach for the day, so I said sure.

When we got to the address, we found the man, Ward Gillespie, at a neighbor's house, where he had been taken in until he could find another apartment. He was sitting in an armchair, and I noticed braces on both his legs. At his feet lay a big German shepherd, who raised his head and regarded us quietly as we introduced ourselves. Ward told us briefly what had happened.

"We don't know how the fire started—the firemen are still investigating. All I know is that I felt Max nudging me and pawing me. By the time I was awake, the room was filling with thick smoke, and I realized I had to get out of there fast. I found my crutches and made it halfway across the

3

room, on my way to the door. The heat was terrible. Max was right beside me, and both of us were coughing like crazy. Suddenly I was aware of falling, and I blacked out.

"When I came to, firemen were carrying me into the fresh air. They had found me on the floor—but in the next room. Max must have dragged me, trying to get me out. The firemen heard Max barking—that's how they knew where to find me. For a few seconds, they were afraid he would try to prevent them from picking me up, that he was guarding me, but they quickly realized Max had been trying to get help. They had to treat both of us for smoke inhalation. There's no doubt that Max saved my life." He patted the dog, who opened his eyes briefly, thumped his tail, and then went back to sleep.

Ward went on to tell us about Max. "I found him four years ago at a filling station—he was kept chained up during the day and shut in the place at night to be on guard. The men treated him badly—he was dirty, thin, and scruffy looking, with some big sores on his back. He was very unfriendly and aloof. I never saw a dog look so dejected and miserable. I just couldn't leave him there. So I offered one of the men fifty dollars for him, and they let me have him." It was hard to believe this handsome, healthy-looking dog stretched out on the floor had come from such hard times.

"Max seems to know my legs are no good," continued Ward. "He is very careful not to get in my way or trip me. When I go up a flight of stairs, he follows, but when I come down, he precedes me, as if to catch me if I should fall. The only thing he doesn't seem to understand is to pick up my crutches if I drop them. When I lose a crutch, and there's nothing handy for me to sit down on, I'm really stuck, be-

4

A disabled person often cannot bend down to retrieve a crutch if he or she drops it. Here, a young trainer is teaching an aide dog to pick up a crutch on command.

cause I can't reach down to the floor without losing my balance."

A light went on in my head. "I could teach Max to do that," I said eagerly. "I'm a dog trainer, and I know how to teach a dog to fetch practically anything. Would you like me to try?"

After Patrick got his pictures of the dog hero and his grateful owner, Ward and I worked out a way for me to come and teach Max for brief periods several times a week. I wish all dogs were as easy to train as that one. Within two weeks he learned to pick a crutch up off the floor or ground and put it in Ward's hand.

I also helped Ward teach Max simple obedience, such as sit, stay, and lie down on command. Ward was proud of Max. He had greatly loved the dog from the moment he got him, and he was very mindful of the fact that his animal had saved his life. But he also seemed to get such pleasure and pride in teaching Max obedience.

Months later when I telephoned Ward to see how he was doing, he told me that the company where he works as a computer programmer lets him bring Max to work now, just as if Max were a guide dog.

The experience with Ward changed my life—I began to train dogs for other handicapped people.

Getting back to Louise and Alfie, I realized that it was unlikely that the child would be able to give a convincing verbal command that the dog would listen to. So I concentrated on teaching Alfie to sit in response to a hand signal from Louise. I once had a student who couldn't speak at all, so I helped him train his dog to obey a little bell.

We usually work in the enclosed yard back of my house where I like to do my teaching in good weather. Each child in these classes has an adult volunteer who helps the child practice with his or her dog. The volunteers are all friends of mine or people who have trained their own dogs with me in regular classes. They get interested in my work with these kids. Sometimes the children's parents want to help, but I've found the volunteers are better. The parents generally try to help too much, instead of making the children try harder. "Can't you do a little more?" my volunteers often ask the children, and usually the kids can.

The first time Alfie sat down in response to Louise's hand signal, the child could hardly believe her eyes. A big surprised smile spread over her face, and she looked up at her volunteer, her face radiant. If I had never before realized what it means to these kids to succeed in controlling their pets, I knew it now. Also, learning to command their dogs helps these children make decisions themselves, rather than rely on other people.

Before the class term was over, Louise could also make Alfie shake hands and roll over, which delighted her and amused Alfie. The only thing I had to give up on with Alfie was getting him to stay as well as sit when Louise told him to. He would sit perfectly, but when Louise rolled away in her wheelchair, that dog simply would not remain where he was—he would get up and trot after her every time. I think Alfie felt protective toward Louise. Well, it was a small failure compared to the successes Louise achieved with Alfie.

My other classes with handicapped children involve high-school kids. I have one class for the physically handicapped, and one for children who are slow learners—that is, intellectually handicapped. And one class is for youngsters who have normal intelligence but for various reasons are doing poorly in school. These three classes are all somewhat career-oriented. The idea is to expose the youngsters to working with dogs in ways that they might consider as job options for the future. In addition to learning dog training, they learn grooming, for instance, and kennel work. They are selected for my courses by their special-education teachers at the high school.

I don't have the kids bring dogs of their own for this training because I want to be as sure as I possibly can be that they will be working with dogs that they can succeed with. These particular high schoolers have had very few

7

achievements in life so far. I want them to know the feeling of working hard at something—and mastering it. So I borrow dogs for these classes from friends and acquaintances, even from local shelters, choosing carefully.

I remember one really tough kid, Mark, a big bullying boy who slouched into class the first day with a scowl on his face.

Oh, wow, I thought. Here comes trouble.

Well, as it turned out, one of the dogs I had borrowed for that class was a lively, powerful young Newfoundland. Maybe that's the dog for Mark, I said to myself. The dog had had some beginning obedience training, but hadn't completed it. Also, he had been at the lake all summer with his family, and I figured he might have forgotten everything he'd been taught.

"Do you think you can handle this dog, Mark?" I asked. "Remember, now, the only way you'll ever get a dog to obey you is by being firm but kind. This dog will require a lot of patience."

I had already stressed to the youngsters that we never, never strike an animal or punish it in any way. Training is always done with firmness and with rewards of praise and petting when the dogs perform correctly. The dogs have to like learning. I kept an eye on Mark to be sure he wasn't rough with the Newfoundland.

Mark did a good job with the dog. He also did a complete 180° turnaround in his personality, at least in class. He became much more cooperative, stopped picking fights and hassling the other kids, and really got into the work with the dogs. His self-esteem seemed to rise significantly. He especially liked grooming the dog—he would comb and

8

brush the animal till its coat gleamed like ebony. It crossed my mind that possibly Mark had never before known how to use his hands gently or learned the pleasure of stroking anything.

Mark showed special interest when we visited a boarding kennel and saw the handlers taking care of the dogs there—exercising and bathing them and the like. He mentioned he might like to go into kennel work. I don't know if Mark ever followed through—I've often hoped he did.

I especially like training dogs to be real aides and companions to handicapped people who are confined to wheelchairs. Besides providing important companionship, the dogs fetch and pick up objects that fall on the floor. I have a list of severely disabled persons waiting for dogs, and I train each one to order, to assure that I match a dog with a person so that both will be happy.

This dog is being taught to fetch and carry for her master, who has little use of his arms and legs. She will be his valued helper and companion.

MARY BLOOM

I prefer golden retrievers for aide dogs, though I've also had a few mixed breeds that worked out beautifully. I like a sturdy dog because some handicapped people use the dog as a support to lean on when they're getting in and out of a chair or bed.

The people at the local pound let me know whenever a young, healthy, suitable dog is available for adoption. Also, a friend of mine who breeds golden retrievers donates a pup now and then that for some reason is not championship quality—for instance, the pup may have a little white marking around the muzzle. However, it occasionally happens that the person I'll be training a dog for wants the pleasure of showing it in dog shows. In that case I try to choose a winner.

Whenever I get a puppy that I will eventually train, I place it in a foster home where it will become socialized and be taught basic obedience while it's growing up. Not long ago, I acquired a delightful puppy that I placed in a foster home in the care of a young girl named Emily, who lives with her family and her own dog. Emily is sixteen years old and in high school, but is already an excellent trainer. When the puppy, Bonnie, was about eight weeks old, Emily started giving her obedience training five or ten minutes every day. Bonnie learned to come, lie down, sit, and stay. She had a little trouble with stay at first—it's hard for a lively young dog—but Emily finally taught her.

By the time Bonnie was eight or nine months old, she would walk to heel on the leash and ignore distractions such as other dogs, cats, birds, and the like. After another few months, she was ready to be trained as an aide dog. I went to pick her up from Emily.

"I hate to give you up, Bonnie," said Emily, hugging

the dog. "I knew from the beginning that I would only have you for a year. But I'll miss you. So will Rover." She patted her own dog, who had come out to the driveway to see what was going on.

"I know you'll miss Bonnie," I agreed. "But think about how much she'll be loved and needed. And remember, you have helped a handicapped person by doing such a fine job raising Bonnie."

"I know." Emily smiled. "I feel good about that. Also, I'm going to write a paper on my experience with Bonnie for English class. My teacher is very interested in hearing about this work."

Bonnie was going to be trained for a young woman named Ann who had suffered head and spinal injuries in an accident and was almost totally paralyzed. When I first met her, she was getting physical therapy daily, but could hardly speak, couldn't sit in her wheelchair for more than two hours at a time, and couldn't move anything but her head and one arm a little. Her husband was devoted to her and urged me to train a dog for her.

"Ann has always loved dogs," he said. "I think it would help her tremendously to have one. And besides, eventually when she's better, she will have to be alone during part of every day while I'm at work. I myself would feel better if she had a dog."

In the few months that followed, I concentrated on teaching Bonnie to pick up whatever I dropped and place it in my hand. I have an old wheelchair that somebody donated to me, and I use it in my teaching. I trained Bonnie to walk beside the chair, with or without her leash.

I also spent a lot of time teaching her to fetch. Out in the field, hunting dogs are taught what's called directed re-

11

trieve—to go bring back birds that have been shot down. I use directed retrieve to get a dog to fetch things from across a room—a shoe, glove, sweater, glasses case, cassette, book, or whatever. Bonnie would go in the direction I pointed and keep on fetching objects to me until she brought what I wanted. After a while she learned certain words too, so that when I said "Fetch slipper," she would go right to the slippers and bring them and nothing else.

One day I pretended to fall out of the wheelchair. Bonnie didn't get upset, because she knew I could get up and walk, but I wanted to see if she would fetch me the telephone.

"Fetch the phone, Bonnie," I mumbled, lying on the floor. Bonnie stood over me, eyeing me with interest for a minute.

"Phone," I repeated. "Go fetch."

Sure enough, she went and picked up the receiver in her mouth. Of course, the stand fell off the table as she brought the receiver to me, but she dragged it along on the cord, and put the receiver in my hand. I reinforced this as Bonnie's training went along, thinking it might be useful to Ann sometime.

Finally Bonnie was ready to go to Ann, though officially she would continue to be my dog for six months. I always retain ownership of the aide dogs I train for this period, to be sure the dog works out and that the person is helped and loves the dog—and also to be sure the dog is happy with the person. Once I trained a dog for a man who, it turned out, didn't really like dogs. I was never sure he actually abused her, but she seemed unhappy when I went to check on her, so I took her back at once. Not everybody is right for an aide dog.

It was really fun to watch the bond grow between Ann and Bonnie as I helped Ann learn to give the commands that Bonnie knew. For one thing, Ann's speech rapidly improved. Apparently she talked a lot to Bonnie when they were alone together.

I think I can guess the reason for this. An animal doesn't mind if your speech is blurred or if you grimace and make faces when you're trying to talk. You are under no pressure to have good diction. A handicapped person who might be embarrassed or easily discouraged trying to make himself or herself clear to another person can practice speaking to a dog without feeling inadequate. Consequently, the handicapped person is bound to improve—and gain self-confidence.

After several months, Ann began to enjoy going out in her wheelchair with Bonnie. Till then, she had always preferred to stay home, even when her husband urged her to let him take her somewhere. Now, they go out together sometimes, and Bonnie almost always goes along. Ann likes the way people admire the beautiful dog walking beside her chair. Naturally, after the trial period, Bonnie became Ann's dog for keeps.

Next thing I heard was that Ann started to do some work at home. It turned out that she is fluent in Spanish and can get work sometimes as a translator. She reads the Spanish text and translates it slowly but clearly into English on a tape recorder. I can't give credit to Bonnie for that, but I do believe the dog plays a part in Ann's journey to rehabilitation. Ann might never improve enough to walk, but she has come far from the sad and silent young woman I first met.

Just as not everyone is meant to have a dog, not all

dogs are suitable to be aide dogs. That doesn't mean there is anything wrong with the dogs. My own dog, Byron, for example, would never make a good aide dog—he just doesn't have the temperament for it, though it would be hard to find a brighter and more loving pet than Byron. That's where my experience as a dog trainer is useful. I can usually judge in advance if a dog will work out well.

One of the keys to success in this work is good judgment—being able to evaluate people and dogs realistically. And just as not every handicapped person is right for a dog, and not all dogs make good aide dogs, similarly not all dog trainers are suited to working with the handicapped. I can't begin to tell you how much patience this work requires. Sometimes handicapped people can only go very slowly. They often have terrible difficulty in carrying out their part in training their dogs. They may not be able to speak clearly or loudly enough. Their arms and hands may be weak, so that the simplest maneuver with the leash is an effort for them. The trainer not only has to be able to understand and control dogs, but must learn a good deal about many different types of handicaps.

Also, you have to be able to innovate, to think of new ways to solve problems. Energy and endurance are necessary too. You don't train dogs, or work with dogs and the disabled, sitting at a desk. You work on your feet.

In training aide dogs for handicapped people, I use everything I've ever learned. And the people I work with have not only taught me much about the problems of the handicapped and how to help them—they've also shown me the best of the human spirit.

14

2

HORSE POWER

Alexandra
describes her work
as a therapeutic
riding instructor.

I saw the sign that read "Oaklands Riding Center" and turned in the driveway. On their business stationery it had said "Therapeutic Equitation" in smaller letters under the title.

I had just graduated from college, where I had majored in physical education and taken equestrian science because of my love of horses and riding. I'd heard about therapeutic equitation, the fancy name for horseback riding for the handicapped. In therapeutic equitation programs, blind people, amputees, cerebral palsy victims, paraplegics, mentally retarded people, and many others with disabilities can have the exercise and pleasure of riding, with the help of professional instructors and specially trained horses.

I wanted to see this firsthand. So, I'd written ahead and gotten permission from Mrs. Anderson, director of Oaklands, to visit and learn something about therapeutic equitation. I parked my car and walked around to the stable. A young woman about my age, dressed in riding clothes, was coming out of the tack room carrying a bridle.

"Hi—I'm Alexandra Snow," I said. "I've come to observe for the day."

"Hello—we're expecting you." The girl smiled in a friendly way. "I'm Cathy, one of the instructors."

Just then a woman in a wheelchair came rolling into the barn. She was severely disabled, but there was nothing weak or timid about the way she pushed herself in that chair.

"Hi, Maria," said Cathy. "We'll bring Silver out for you in a minute. Boy, does he need brushing. We'd like you to do a really good job on him today."

Brush him! I thought with dismay. Do they seriously expect that woman to groom a horse? I doubt if she can even stand up.

I was in for a surprise. Cathy led Silver out of his stall and casually handed the reins of his bridle over to Maria. Leading the horse with one hand, Maria wheeled herself out of the barn. I tagged along and leaned against the fence to watch.

In front of the stable, Cathy held Silver's bridle while Maria wheeled her chair right up beside him. Most horses would have flinched at that, but Silver didn't bat an eye. Then, Cathy handed Maria a brush. Maria struggled to her feet and brushed the horse thoroughly on one side. It took a lot of effort on her part. Then Cathy had the disabled woman go around and brush the other side. I thought Cathy was sort of hard on Maria, but Maria just laughed and did

16

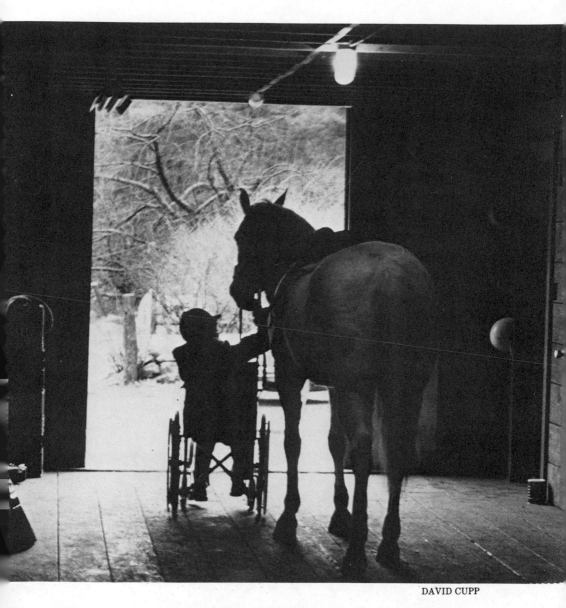

At a center for therapeutic horsemanship, a victim of cerebral palsy leads a well-trained horse from the stable into the yard.

as she was told. Cathy did help her put the saddle on, at least.

Another instructor, whose name was Bill, soon joined them, and he and Cathy teased Maria a lot. Maria kidded them right back. I got the impression that beneath all the kidding around, Cathy and Bill respected Maria greatly.

Then, the two boosted the handicapped woman into the saddle. Silver just braced his feet and stood perfectly still as Maria's weight landed awkwardly on his back. Maria gathered up the reins and, with Cathy walking beside her, rode down the hill to the riding ring.

I could hardly believe my eyes when I saw Maria ride at a trot. Her head bobbed a good deal, but she managed to post, pushing herself up and down in the stirrups. Cathy ran, leading Silver, and someone else ran alongside just in case Maria started to fall, which she didn't. Silver acted as

To see this woman ride, you would not guess that she cannot walk. An instructor leads the horse at a trot; another runs alongside in case the rider should need help in keeping her balance.

DAVID CUPP

if he knew he should pick up his feet carefully and not stumble, swerve, or break his stride.

"It's thrilling," Maria told me later. "I was really scared when I first came here to ride. Can you imagine what it felt like to someone who had never even walked, to be lifted onto a *horse*? I was in very poor physical shape, could not push my wheelchair myself, and couldn't even raise my arms, much less stand.

"But now I've been riding nearly three years, and it has made such a difference in my balancing ability. The motion of the horse, the need to always correct my balance in the saddle, has given me exercise I'd never had in the wheelchair, and it has helped strengthen my whole body. Now I've learned to walk with help, and I can use crutches. Did you see me stand in the stirrups and post when Silver trotted?"

I told her I certainly had seen. "You looked great," I assured her, and it was true.

"But most of all," Maria continued, "being able to do this gave me confidence in myself. I had always lived with my parents, and they took care of me. Now, I have my own apartment with a roommate, and I have a part-time job. My life changed when I learned to ride."

Flashback to five or six years earlier, when I was in high school. I had my own horse; his name was Blaze. I took complete care of him myself, including buying his feed and paying for his routine veterinary care with the money I earned babysitting and working part-time at the supermarket. I rode him every day. We would trot slowly past the neighbors' houses till we came to a smooth dirt road that stretched for several miles, with a good view ahead, no hidden driveways, and virtually no traffic. That was where we

19

always cantered. When I was younger, I used to pretend I was a jockey riding in the Kentucky Derby. (I had to give up that fantasy when I continued to grow—did you ever hear of a 5-foot-8-inch, 120-pound jockey?)

One fine autumn day we were walking past a house on our way home, and I happened to glance at some people beside a car in the driveway. They were unfolding a wheelchair, while a man lifted a girl about my own age out of the car. She was a really pretty blonde girl, but pale and fragile looking. I saw her legs were withered and in braces. The man set her in the wheelchair, and just then she looked across the lawn at me. Blaze was prancing a little, exhilarated by the good gallop we'd just had, and I was controlling him easily because I knew him so well. The girl watched us, and our eyes met. I'll never forget the way she looked at me. Something very strong stirred inside me. A lot of mixed feelings pushed their way into my mind—pity for her, pride in myself and my horse, and at the same time a lot of guilt. Why should I be able to do this marvelous thing, while she could not?

Then Blaze and I passed on down the road, but I thought long and hard on my way home. An idea began to form in the back of my mind. Okay, the girl couldn't walk, but did that necessarily mean she couldn't ride? Would it hurt her if she did? Would she be scared? Couldn't somebody walk or run beside her and hold her on the horse's back? Had anybody ever tried it?

Although I rode past the house many times after that, I never saw that girl in the wheelchair again. I didn't forget her, though.

I thought of becoming a veterinarian when I grew up,

so I could be around horses. I figured that if I really knocked myself out in college and got spectacular grades in the preveterinary courses, I might possibly get into veterinary school and specialize in large-animal medicine. But later on, I discovered a field of college study that I liked even better. I had always loved sports—baseball, basketball, soccer, swimming, skiing, and best of all, riding. I decided to major in physical education in college, and take equestrian science.

College was a pleasure; the four years flew by. I wasn't sure exactly what sort of job I'd look for when I graduated, but I felt I was learning skills that would be useful to me in some way.

The idea of teaching riding to the handicapped interested me. You can imagine how I felt when I saw Maria at Oaklands that day. Here was someone like the girl I'd seen that afternoon, years before, with Blaze—and she was riding, just as I had wished that other girl could.

Maria wasn't the only physically handicapped person I talked to at Oaklands. There was also a tall, tanned man who told me he was paralyzed on one side of his chest as a result of a spinal injury suffered in a skiing accident.

"I thought my athletic life was over," he said. "I was very weak, couldn't walk much, got out of breath quickly. When I first came here to ride a year ago, just getting on the horse was an enormous effort. Now I ride once or twice every week. My lung expansion is much better—I don't get winded anywhere near as easily." This man wore a neck brace, but he had ridden briskly in the ring without any help at all. He rode a big easygoing horse who was much like Silver in temperament.

21

"Most important is the way I feel—the mental uplift is tremendous." He smiled. "I can go trail riding and enjoy the outdoors again."

As I listened to him and Maria, and watched Cathy and the others work with the horses and riders with such care and concern, I began to think, Maybe I could do this.

I was talking to the director, Mrs. Anderson, when a school bus drove into the stable yard and about thirty children ranging in age from perhaps eight to twelve got out. They were lively and healthy looking, and certainly seemed bright enough.

"Who are these kids?" I asked.

"They're from St. Joseph's, a home and school for emotionally disturbed children," Mrs. Anderson told me. "They're at that school because they have really been through the mill—they've been abused, neglected, abandoned. Not surprisingly, they have a lot of problems. They come here as part of their therapy." And she went over to take charge.

Supervised by their counselors or Oaklands instructors, the children went into the horse stalls in groups of two or three and proceeded to comb and brush the horses and then saddle and bridle them. Except for a few kids who hung back, most of them went about the tasks with rapt attention.

"The animals capture and hold the children's interest, which in itself is therapeutic. These youngsters have a short attention span, and they're usually unable to concentrate," Mrs. Anderson explained to me. "Also, mastering their initial fear of such a large animal helps develop their self-esteem. They develop memory and self-control. To children who are accustomed to failing at virtually everything,

22

Alexandra, all this is important. Coming here and being with the horses helps them in many ways."

After the kids finished in the stalls, they performed calisthenics and tricks on one of the horses that wore a vaulting surcingle—a strap with handles on top, the kind used by gymnasts and circus acrobatic riders. With Mrs. Anderson close at hand, each child, one at a time, ran to the horse and was boosted aboard. The child did tricks and exercises such as standing up on the horse's back, standing on one foot, turning around completely, reaching up for the overhead beam of the barn, lying on the horse's back, and the like.

"I can't, I can't!" cried one little boy each time Mrs. Anderson asked him to perform an exercise in the saddle. And every time, with her calm, reassuring voice encouraging him, he went ahead and did the exercise. And whenever he did it right, he broke out into a radiant smile of pride.

Throughout all this, the horse stood tolerantly. She simply braced herself as each child scrambled onto her back and then stood almost unmoving while the youngster performed. In the stalls, the horses had been quiet and gentle while the children groomed them and milled around them. Looking back at this scene in my memory, and having observed so many similar scenes since then, I could swear the horses understand. Is it possible? People think horses don't have the brains for abstract thought, but they certainly seem to make allowances for the disabled people I've seen them work with. They seem to pick up what's going on—perhaps by some means of understanding that we haven't yet discovered.

"In a few months, I want every child in this group to be able to get on a horse by himself or herself, and be able

23

to hold the reins properly," commented Mrs. Anderson to me. "At that point, we'll evaluate each youngster in terms of riding in the ring, and start by leading each rider around. Make no mistake, Alexandra, this work requires almost unlimited patience."

I was greatly impressed by the day I spent with Mrs. Anderson. She was very inspiring, and also I liked the fact that horses were used this way as full partners in therapy for people who needed them. I did some homework to find out more about therapeutic equitation. I learned it's not a brand-new idea—there are references in the earliest medical writings, from ancient times, to the use of horses in the treatment of certain disabilities. As medicine became more of a science, therapeutic riding fell into disuse—I suppose because it didn't seem scientific.

In our time, one of the first people to explore the benefits of this sport for handicapped people was a determined Danish horsewoman, Liz Hartel, who caught polio and became crippled, confined to a wheelchair. Mrs. Hartel was not about to give up her riding. She insisted on resuming it, struggled hard to regain her strength and competence, and won the admiration of the whole world by competing in the 1952 Olympics and winning a medal!

The following year, Pony Riding for the Disabled, the first school built solely for the purpose of teaching riding to the handicapped, was founded in England. Its director, John A. Davies, developed the instruction methods that are still generally used today.

In 1970, the Cheff Center for the Handicapped, the first institution in North America specifically built and staffed for teaching riding to disabled people, opened in Augusta, Michigan. Its director, Lida McCowan, a lifelong horsewoman, had trained for this work in England. Some two hundred thirty disabled adults and children ride at Cheff every week.

Today, there are about two hundred centers in the United States and Canada that are accredited by the professional organization, North American Riding for the Handicapped. This group works to standardize teaching methods and qualifications for instructors. It also evaluates the program quality and safety standards of places that offer riding to the handicapped, and accredits those that meet certain qualifications.

Cheff offers a three-month intensive program to train instructors. After visiting Mrs. Anderson's center, I applied to that. Since I had my degree in phys ed and had been riding almost all my life, I thought the course would be easy for me, but it was tough. First of all, only 35 percent of all applicants are accepted—when I learned that, I realized how lucky I was. Students apply from all over the country. An average of three out of four graduate.

We were taught routine horse care and stable management. We also learned human anatomy, orthopedics, and physical therapy, and much about the various disabilities we would be likely to encounter in our work, such as cerebral palsy, spina bifida, and paraplegia. Physically handicapped people are the ones who benefit most from riding therapy. One of my favorite riders at Cheff was a four-year-old boy with no hands. He held the reins with the hooks on the ends of his arms and rode wearing a great big smile. But emotionally disturbed children and retarded people gain confidence, self-control, and self-esteem that cannot be underes-

25

timated. And it is not unusual for autistic or other severely withdrawn children to start speaking for the first time during or after a ride. One child had her first real, spontaneous conversation with one of the Cheff horses. She had disappeared into the barn when nobody was looking, and we found her sitting on the hay in a stall talking to the horse.

The twenty-six horses and ponies at Cheff were wonderful animals and, like the horses I saw at Oaklands, seemed to understand so much. One day something happened that would have convinced many disbelievers of the sensitivity of horses.

There was a small group of retarded young teenagers riding in the ring. These kids had been riding for some months and were pretty competent, so Sandy, the instructor, decided to let them canter one at a time. Two of the youngsters had done very nicely, but the third child lost her balance and fell off into the deep, soft sod of the arena. Nobody has ever been injured at Cheff, but this girl slid forward over the horse's shoulder and landed right under one of his front feet. The horse froze, with his foot lifted. He stood that way the whole six seconds or so that it took us to reach the girl and help her up. Then the animal put his foot down. Now, I ask you—why would a horse be so cautious unless he understood that the people who rode him needed special care?

I'll never forget all I learned at Cheff—about horses and teaching, and about the disabled.

So—here I am now, working as a therapeutic riding instructor at Green Haven, a residential school for retarded children. It doesn't look as if I'll be rich and famous any time soon, but I couldn't be happier. Every day, unless the weather is too wet or cold, kids come down to the barn for

26

their riding lessons. The children are all ages, physically, but they range in mental age from under six months (the lowest on the scale—really profoundly retarded) to perhaps six or seven years old (highest on the scale—quite responsible). Some have physical as well as mental disabilities. And all of them benefit from what the other instructors and I teach them.

At first, I had a hard time getting used to some of the children, who were in worse shape than those I had worked with at Cheff. Some look grotesque, can't walk straight, can't speak plainly, maybe drool, sometimes go into uncontrolled rages. And I was dismayed when I saw a few of the full-sized twelve- or fourteen-year-olds who are unable to even get out of bed, speak, or recognize what's going on around them. But now I am completely used to all of them, and have grown to love many. These kids are very endearing, and I get a thrill when they respond to riding and it helps them.

Take Rhoda, for instance. She wouldn't come near the barn for almost the entire first year she lived at Green Haven. She would scream and try to run away every time a counselor approached the barn with her. After a long time, she got so she would come to the riding ring and watch a little. Then, holding tight to a counselor's hand, she would walk up to a horse and inspect it. But she would stiffen in terror and cry if we tried to lift her onto one of the ponies.

Now you should see her. When Rhoda is in the ring, she can't wait for us to take the horse off the lunge line so she can steer the animal around herself. She sits up straight in the saddle looking confident and pleased with herself. And since she has been riding, she does better at everything else, I'm told.

27

Retarded children benefit greatly from horseback riding—it gives them a sense of control and achievement. Here, a proud and happy little girl competently handles a gentle horse with hardly any help from her sidewalker.

Then there are children like poor Bryan, a profoundly retarded boy who can't even sit up by himself. We lay him across the saddle on his belly, and one of us holds onto him while another instructor leads the horse around at a walk.

We aren't sure how much Bryan gets from the experience. We believe the smell, warmth, sound, texture, and motion of the animal give him some sensory stimulation. We do know the movements of the horse give him some exercise. He tries to hold his head up while he's across the saddle, and this strengthens his neck muscles and gives us hope that some part of his brain is responding. Our immediate goal is for him to improve enough so that we can hold him upright in the saddle.

Each child's riding program is individually tailored to the child's need and potential. Every youngster always wears a helmet and a belt with a handle at the back for the instructor to hold onto if necessary—no one, no matter how helpless, is ever tied onto a horse. Sometimes one of us rides as a support rider, holding the child in front of us in the saddle.

First we teach the children how to balance themselves on the horses' backs; they do stretching and reaching exercises in the saddle that make them correct their balance continually. We don't let them hold the reins until they have mastered that—if they kept pulling on the reins to balance themselves, that would be too hard on the horses.

Then we lead them over uneven ground to help them improve their balance and coordination further. Some children have such poor muscle tone that they have never been able to stand up alone, but they actually may learn to walk after they have been riding for a while. Also, the continual adjusting of their balance in the saddle stimulates the vestibular or balancing part of their inner ears, which has never been developed because they have been sitting in wheelchairs all their lives.

We also pay a great deal of attention to the horses and ponies. Their health and comfort are extremely important. After all, these animals are in a way our cotherapists, and they need—and deserve—the best of care.

Several times a week, after the children have ridden, the other instructors and I ride. It would not be good for a horse to be ridden only by retarded people. The reactions of even retarded children who can ride are much slower than those of normal persons. Also, the most severely retarded children are just dead weight on the horses, and this overtires the animals. So the horses must be ridden by others frequently to keep them from going stale.

One of the most important parts of my job is training horses for this work.

"Another horse has been offered to us—a seven-year-old mare," said Mrs. Fuller one morning. Mrs. Fuller is the senior staff member in charge of the riding program. "She'll be here tomorrow. Alex, will you make time to work with her for the next few weeks and let me know what you think of her?"

The horse belonged to a family with children who were now too busy in high school and college to pay attention to their pet. Green Haven often receives horses donated to us for this reason. Mrs. Fuller had questioned the owners at length, and this horse sounded like just the sort we can use. But the animal would still have to go through a three-month trial period before we accepted her permanently. As you can imagine, we have to be extremely particular about our horses and ponies, for both their sake and the children's. Many horses turn out to be unsuitable for use in therapeutic equitation.

I was just coming out of the stable when the trailer with the new horse pulled up the next day.

"Ah, you're a beauty," I said to the mare as she was led down the trailer's ramp. She was chestnut with two white hind legs and a star on her forehead. I learned her name was Sarah. She seemed very gentle and not especially upset about the trip or her new surroundings. I led her to a stall in the stable to rest up from the trip before I rode her. Late the next afternoon, when all the therapy riding was finished for the day, I saddled her and rode into the ring. In a way, she reminded me of Blaze—her gaits were similar. (When I had gone off to college, I had sold Blaze to a family that live near my parents. They love him, and I still visit him whenever I go home.)

Over the next three months, I rode Sarah every day. Sometimes I subjected her to the kinds of unpleasant treatment she might occasionally get from the children. I flapped the reins, kicked her in the ribs, shouted and screamed, wiggled around in the saddle or just went limp and bounced. I had one of the other instructors throw a ball in front of Sarah while I was in the saddle, to see what she would do. Sarah jerked her head back out of the path of the ball, but she didn't jump or bolt. I rewarded her with petting, praise, and treats of apple or carrot when she endured these irritations without flinching. She soon got the idea that what I wanted was for her to accept uncomfortable treatment calmly.

31

Sarah came through her trial period with flying colors. Her only fault was that she tended to be a little lazy. A child might have to urge her into a trot—Sarah preferred to amble around the ring. She worked well enough on the lunge and tolerated everything, but she did always seem relieved to go back to the barn. She was so sweet, however, that we forgave her for being a little dreamy.

Our star rider is a great kid named Joey, a real character. Joey is about twelve physically and maybe six mentally, active and cheerful and eager to try everything. But it wasn't always so.

Mrs. Fuller told me a wonderful story about Joey. When he first came to Green Haven, he was a very withdrawn child. When the counselors brought him for horseback riding, he would meekly let himself be put on a horse and led around, but he didn't show any emotion one way or the other. He never responded to affection, either. The staff here is very caring. The kids get a lot of physical contact—hugging, patting, stroking, that sort of thing—and in turn most of them are extremely affectionate. But not Joey, not in those days.

Then one day when he was hanging around the horses and ponies waiting his turn to ride, he went up to Willing, a wonderful big, gentle horse who's just like his name. Joey began to put his fingers in Willing's nostrils and mouth. One of the instructors started to interfere—afraid not that Willing would bite but that Joey would annoy him. Just then Joey laid his cheek tenderly against the horse's warm velvet nose and stood there like that for a few minutes.

"Nice Willing, nice Willing," the boy said softly.

The instructor stood back—she knew she was witnessing a real breakthrough. In fact, no one had been aware that Joey even knew the horse's name.

Joey began to improve in every way after that. Now he is so sunny and outgoing, it's hard for me to picture him as the sad and passive child he once was.

In this business, we get rewards like that every so often. People taking care of retarded children work hard to keep them from being further dulled by continual failure. Being helped to succeed at something—especially something as exciting as riding a big animal—does wonders for their self-esteem and gives them confidence to try harder at the tasks given them. We set realistic goals for the children and take comfort in their small achievements, while at the same time we always try for a little more. Many of the kids work hard. If you think about where they're coming from, you can fully appreciate their accomplishments.

It would be a mistake for us to assume that therapeutic riding can help every type of physical disability. Nor would it be possible to use it with severely psychotic people. But its potential has not yet been fully explored and documented. I would like to do that. I intend eventually to get my master's degree and even perhaps a doctorate in physical therapy and go on to expand this field.

Meanwhile, I am outdoors and active every day, and I consider myself extremely lucky to be involved in work that keeps me happy and interested. Not only am I around horses all the time, but my work is useful and needed. To see a handicapped child struggle and then finally accomplish a goal, and see so plainly what this means to that child—this makes me feel good. To me, there couldn't be any work that's more rewarding.

33

3

EYES WITH FOUR LEGS

Bob, who speaks here,
is an apprentice trainer
of guide dogs
for the blind.

If you saw Gretchen walking briskly down the street with her head held high and her black hair flying behind her, the beautiful golden dog pacing beside her, you would probably think she was just a great-looking young woman out walking her dog. Then you would notice the harness on the dog, the traditional harness with the tall handle. Astonished, you would take a closer look at Gretchen—could this confident person really be blind?

She was my student. "Wait, Gretchen," I called. "Slow down a little. You're getting too far ahead of us."

As far as I was concerned, there was just one problem with Gretchen—she was too independent. She made me nervous. Usually, blind students are reluctant to trust their new dogs. They have to be helped to rely on the animals so that they can become more independent—after all, that's

34

the main point of guide dogs in the first place. So Gretchen's faith in her dog, Sunny, was really *my* problem, not hers.

I'm an apprentice instructor at a center where guide dogs are trained. Today, I was working in town with a group of students and their newly trained dogs so they could get some exposure to crowds and traffic together. It was a broiling summer day, and the dogs must have been uncomfortable walking on the hot sidewalks, but they were enduring it and doing their job pretty well. The blind students weren't complaining. The instructors were calm. I guessed I was the only one who felt tense, though I hoped I wasn't showing it.

This was the first time I had worked elsewhere than on the protected grounds of the training center or in the quiet suburb near us. Here, there were so many people hurrying along the sidewalks; cars, trucks, and buses speeding in the streets; noise and confusion everywhere. If I hadn't done a good job with Sunny and the other dogs, I kept thinking, one of them might make a mistake, and Gretchen or another blind student might get hurt or killed, and it would be my fault. I felt totally responsible.

Just then Gretchen put her hand out toward me and touched my arm. "Don't worry, Bob," she said. "Everything's fine."

I was startled. "How did you know what I was thinking?"

Gretchen laughed. "I picked up your nervousness in your voice and body language. Sunny is a great dog. I know I'll be okay with her."

Twenty-three-year-old Gretchen, blind since birth, had had a guide dog before. When he had grown too old to work any more, she had given him to a friend who wanted him and was going to keep him as a pet for the rest of his life. Sunny was her second dog.

"Each dog is different," Gretchen told me later. "When I first started to walk with Charlie, my first dog, I was rather timid. I was used to getting around with a cane and with human helpers. I was very uncertain with him. But Charlie was a very take-charge sort of dog, and he just took over. He knew exactly what to do, and he taught me to follow him. He also made it clear when I was supposed to tell him what to do. He'd stop and wait with an 'Up to you, now, Gretchen,' way about him. He understood that we were partners. Sunny is a very competent dog too, though not as strong a leader as Charlie was. But then, I don't need to be taught so much this time."

Gretchen was the bravest girl I'd ever met—I really admired the way she lived her life. She just acted perfectly natural. She had always gone to regular schools, not special ones for handicapped children, and now she was about to get her master's degree at college. I was grateful that she was trying to reassure me. Yet, it wasn't the first time I'd been helped by the students. Sometimes it's a case of the blind leading us greenhorns.

I am in the third and last year of my apprenticeship in guide-dog work. If all goes well, I will earn my certificate as an instructor, qualified to train guide dogs and blind students. And after five or six years' experience, I will be eligible to train other instructors.

How did I happen to go into this work? Well, after high school I didn't know what I wanted to do, so I joined the army for two years. Because I've always loved dogs and had dogs of my own, I volunteered for the K-9 Corps. When I got out, I was an experienced trainer of dogs for certain kinds of work such as sentry, patrol, drug and explosive detection, and other skills useful to the army.

So at first I thought of becoming a professional trainer of guard dogs, but I was uncomfortable with the idea of turning dogs into potential weapons. It certainly wasn't what I wanted to do in civilian life. Many of the schools that train dogs for that work brutalize the animals in order to make them vicious—and that just isn't my cup of tea. I could have simply trained dogs professionally for obedience or for behavior in the show ring, but that seemed boring. Then one day I happened to see a young guy about my age on the street teaching a blind person and a guide dog, and I decided to look into this field.

I came out here to this training center one day and after talking to the director, got permission to watch some of the work. A new class of blind students was working with the dogs on the grounds of the center. An obstacle course of poles, tin drums, and other objects had been set up, and the dogs were leading their people around them.

"Go on, June, follow your dog," called one of the instructors to a woman who had stopped her dog in confusion when it tried to lead her up an incline in the path. As I watched, I realized that the dogs knew what they were doing—the blind students were being taught to trust them and give them the right signals.

That day a middle-aged blind man working with his dog caught my interest. He had never had a dog before, I was told. The impact of the animal on him was plain to see. He was fascinated, yet at the same time hesitant about trusting

37

the animal. He was having a hard time walking through the obstacle course with his dog; he was tense and nervous and jerking the harness. He got more and more frustrated. Finally the instructor drew the man aside to a bench and asked him if he'd like to rest for a few minutes. The man sat down looking totally defeated and said he was afraid he couldn't go on, he'd never be able to learn to get around with the dog. Just then the dog, who had been standing patiently beside him, put his front paws on the man's lap, licked his face, and began to paw at his arm, as if to say, "Come on, we can do it."

The man wept. The instructors pretended not to notice, so as not to embarrass him. After a few minutes, the man got up and went through the obstacle course with his dog without any problem.

I was simply knocked out by the whole scene. I got an inkling of what these dogs must mean to these blind men and women. The people work hard to overcome any doubts they might have about entrusting the dogs with their lives, because they are so eager to have the kind of independence the animals can give them. I began to think this might be really interesting and meaningful work.

People come into this teaching field from a variety of backgrounds. Almost all of them are attracted initially, just as I was, more from a desire to work with dogs than an interest in the handicapped. The love of helping people comes later.

I believe most agencies that train guide dogs prefer applicants with at least two years of college, or better yet, with training as veterinary technicians. However, any experience in training dogs, such as mine in the K-9, is good too. I know one apprentice here who had been a kennelman and handler for a breeder. Another had been raised on a farm and therefore was very familiar with animals, and had trained her own dogs in the 4-H club obedience classes.

My friend Debby, who I think is one of the best apprentices among us, has only been training here for a few months but seems like a natural. She really understands dogs and has a perfect combination of patience, firmness, and kindness.

"I trained my first dog when I was fourteen—my parents let me have the dog on the condition that it be very well behaved," Debby told me. "I went to obedience classes with my dog. We lived in an apartment building, and when some of the neighbors noticed how obedient my dog was, they asked me to train theirs, and I did. One summer I worked for a veterinarian. I've always loved being around animals. I spent a year in college studying accounting and then a year in secretarial work before I realized that was the wrong career for me and got onto the right track of what I really love to do."

Experience with dogs, while it is certainly a priority for this work, isn't the only requirement. Physical fitness is very important. As one who has walked thousands of miles in the last two years, ten to fifteen miles a day on the average, I can say that you have to be in pretty good shape. Debby is small and slim, but apparently tough, because she seems to work tirelessly.

Surprisingly, verbal skills count a lot in this work. I never thought I had any, but you do have to be articulate, able to communicate not only to the dogs but to the blind students. Many of them have no previous knowledge of

39

dogs. You have to explain a lot. After all, you are often passing on trained dogs to real amateurs, and you had better be sure the people learn enough from you to treat their animals right and use them to best advantage for both themselves and the dogs. Sometimes we even have to persuade a student to drop out and give up the idea of having a guide dog, if we feel he or she just isn't right for a dog.

I knew one apprentice who related well to the dogs but who quit training because he was so uncomfortable around the blind people. Over half the apprentices, by the way, drop out at some point during the three years. They either quit because of the long hours, pressure, and responsibility, or they are asked to leave because they aren't suited to this work.

So after visiting that day, I applied and eventually was accepted as an apprentice here. The first thing I had to do was forget everything I had learned in the K-9, except the obedience work. I spent my first few months teaching a string of about eight dogs the traditional obedience commands: come, sit, stay, lie down. I drilled them in hand signals for forward, halt, left, and right, and to reverse by making two right turns. I taught them to stop at curbs. All this work was done with the dogs on the leash, not in harness. And oh yes—I taught them to disregard squirrels, cats, other dogs, and to concentrate on working when on the leash. I never used punishment, but rewarded them with praise and petting when they did the right thing. The instructors recommend that we always speak softly to the dogs. They say it keeps them on their toes—the animals have to pay attention in order to catch what we say to them, and they are less likely to get distracted.

Then the dogs I had trained went on to a more experienced trainer, and I got another eight dogs. These I brought further along in their training, including beginning harness work.

Each dog trains with us for about six months before it meets its future owner. The blind people come here to the center and live as students in the dormitory for nearly a month, working with the dogs that have been chosen for them.

All our education as guide dog trainers and instructors, by the way, is on-the-job. We have no classroom work. That suits me fine—I have always learned more from doing than from listening to lectures.

After I had been here a few months, I was given one assignment that served as my initiation into the world of the blind. I had to spend twenty-four hours blindfolded. You might think that would be sort of a game, not too difficult, especially in familiar surroundings. I can tell you it was no joke. And if you think knowing your way around a place helps you when you've been blindfolded for a few hours, you're wrong. The best words I can think of to describe that part of my training are "scary" and "frustrating."

I lost my balance and fell down a lot. I had trouble feeding myself at meals, dropping food all over my shirt. I got lost in the dorm where I stayed with the blind students. Actually, they were a lot of help to me. They tried to teach me to use my hearing, to judge distances and orient myself by sounds, to identify things by the feel of them, and so on. Also, they were cheerful and sympathetic; they kept my spirits up.

I even worked with a dog while blindfolded. The dog

had already been trained to the harness, and a trainer came alongside me. It was mind-blowing to be led blindfolded by a guide dog, and to feel what blind people must feel. I'll never forget that experience.

Let me tell you something about our dogs. While a few have been donated to us, most of those we use are bred specifically for this work. Many agencies that train guide dogs use German shepherds, and we have some of them too. We use mainly golden retrievers and Labradors that we breed ourselves. We also like golden Labradors, a crossbreed of golden retriever and yellow Labrador retriever. They are intelligent, steady, and good-natured, and their size and stamina are perfect.

But it would be a mistake to raise the dogs in kennels at the center, because they are going to be living with people, and should be socialized from birth to human family life. So all except the dogs that are currently in training live in foster homes in the community.

The puppies are born in foster homes and stay with their mothers until they are six weeks old. Then each pup is placed in a different foster home. The families are chosen carefully to be sure they can give the animals the very best start in life. We make sure the puppies are well treated and loved so they will be gentle and friendly when they come back to us after a year, for training. The young dogs should have reliable, stable dispositions, be good with other animals and children; they should not be too boisterous, and they should be loyal but not overly protective. We stress that we are training guide, not guard, dogs. And of course, the young dogs must be super healthy.

Our puppy program works well, and about 75 to 80 percent of the dogs returned to us are trained successfully. For the first two to four weeks they are at the center, the dogs are tested to see if they will make good guide dogs. Dogs that clearly won't make it as guide dogs are returned to the families that raised them and live as pets.

About halfway through their training, the dogs are spayed or neutered, except perhaps one or two that are singled out for breeding. We breed the bitches when they are two years old, and they have only four litters so they aren't exhausted by continual breeding. In order to avoid any poor hereditary traits, we are careful not to inbreed.

Once the dogs are ready for the harness, the training becomes intensive, and this is where the real professionalism of the instructors comes in. It's a far cry from obedience classes. We have to teach the dogs to handle traffic, crowds, transportation vehicles of all types, and places such as stores, restaurants, and other public buildings. We have to teach the dogs to accommodate for the height and width of the persons with them, as well as their own.

And here's the hardest part—we have to teach the dogs on the one hand to obey us, and on the other hand to disobey us if we give an order that's dangerous! If an owner gives the dog a "forward" command crossing the street, and a car is coming, the dog must refuse. The pressure on us is great, and we in turn put pressure on the dogs, because they must be ready on time for each class of blind students. It is hard work, long hours, but exhilarating too. The intelligence of these dogs seems limitless. I couldn't believe the things we taught them, that first year of my apprenticeship.

About sixty dogs each year graduate from our training program and leave to serve blind people.

One day the first class of blind people that I had ever worked with arrived. I was assigned to help one student, a young guy about my age named Michael. I had to get over my initial awkwardness in talking to Michael. My first instinct was to speak up loud and clear to him.

Michael put a stop to that right away. "Hey, Bob, I'm not deaf," he said. "Just talk normally."

Michael already walked with a good firm step, but I had to help him get used to the sudden stops and swerves he might have with a dog. I tried to simulate these by playing the role of the dog. I walked beside him holding onto the harness while Michael held the handle. I taught him all the commands he would use with a dog. After a few days, he was ready. A dog named Sabrina had been selected for him.

Now I not only had to teach Michael everything he would need to know in order to use his guide dog, but help him take over the dog. Sabrina, of course, was accustomed to obeying and leading the instructors. It was essential to transfer her allegiance to her new owner. Michael fed, groomed, and exercised Sabrina every day. She slept in his room in the dorm at night.

A lot of thought and care goes into matching the dogs with the students in terms of size, weight, and temperament. While most of our dogs are somewhat standard in size, we wouldn't give a more lightly built dog to a big, heavy, strong person, and vice versa. A rather sedate dog will go to an older person, while a more spirited and active animal will go to someone like Gretchen.

Our selection committee screens applicants carefully for their health, character, and motivation, and to be sure they can provide well for the dogs. Gretchen represents the best

44

A blind student is learning to trust his four-legged partner. The trainer is teaching him to step up on the curb when his guide dog stops.

45

type of person for a guide dog. At our agency, applicants for dogs must be at least sixteen years old, but age requirements may vary from place to place. There's no top age limit—it depends on the person. We had a vigorous, active, eighty-year-old blind man here last year who was given his seventh dog.

Applicants for guide dogs tend to be highly motivated people who want the animals in order to increase their mobility and independence. Almost all of them are employed. Some, such as housewives, work at home. Some are students. All recipients of our dogs sign agreements that they will never beg with their dogs, and we try to enforce this.

Some agencies that train guide dogs for the blind charge for the dogs. Other places are funded so the dogs can be given free.

When the dogs leave our training center and go home with their new owners, we follow up with personal visits within the following two weeks to see if everything is going well and to help the blind persons work out any problems that may have shown up with the dogs in their new homes. We can't do this with foreign clients, but we do travel almost anywhere in the United States to give this aftercare service. Then, after that, the owners are asked to communicate with us twice a year to let us know how they and their dogs are doing.

One blind young guy, Ted, told us a great story about his dog Bruce. This guide dog, like most others when they are out of harness, loved to romp and play like any dog. Bruce especially liked to play ball; his favorite sport was to have someone throw a ball for him to retrieve. He also enjoyed it when members of his owner's family, who were sighted, played catch—he would give chase as the ball was thrown from person to person.

Ted participated in these games sometimes—the others would throw the ball into his cupped hands, and he would then throw it toward the voice of the next person. But naturally, sometimes when the ball was thrown to him, he'd miss. Then, Bruce would show great concern, and, taking Ted by the sleeve, would lead him to where the ball had landed. He didn't do this when anybody else missed the ball. Shows you how much these animals understand, doesn't it.

So now you can see that training guide dogs is no business for amateurs; it is demanding, professional work. That's why I felt so responsible for Gretchen and the others, the other day in town.

Three students with their guide dogs are practicing walking in a busy downtown area, under the watchful eyes of their two trainers.

MARY BLOOM

One thing Gretchen and Sunny had to do that afternoon was go in and out of the subway. Riding the trains was not the first thing Sunny had to learn—she had to take Gretchen through the turnstile. The dogs, of course, can walk right under, but they must learn to stop so they don't lead their owners smack against the turnstile. After Gretchen and Sunny went through several times, we were satisfied that Sunny was catching on. And she wasn't nervous at all when a train came roaring into the station. Several days later, we came back and tried again. Sunny stopped at the turnstile. And she led Gretchen onto the train with perfect poise and rode calmly to the next station.

In order to know when to get off the trains, by the way, blind persons count the stops to their destinations. Many say their dogs learn the stops also, and are on their feet, ready to get off, before their owners give them any signal. Do the dogs have their own ways of identifying the stations—ways that human beings haven't discovered yet?

Now my students have begun to make their first solo trips away from the training center. I felt a pang the first time they went off, and paced up and down as the time approached for them to return. The senior instructors teased me about acting like a mother hen.

One hot afternoon after working in the city, Gretchen and I decided to have a soda in a coffee shop and wait for the others. A couple in the next booth were fascinated by Gretchen and Sunny. The man reached out to pat Sunny on the head.

"Don't do that!" his companion cried. "You're not supposed to pet guide dogs."

"Oh," said the man. "I'm sorry—I didn't realize. I should have asked you first."

"That's okay," said Gretchen. "I don't mind your petting her when we're relaxed like this. It's not a good idea to touch a guide dog while it's working, though. The dog has a lot on its mind. It has to pay attention, and you might unintentionally distract it if you try to pet it or talk to it."

"Please tell me," the woman asked Gretchen, "when I see a blind person with a dog standing on the corner waiting to cross the street, I never know if I'm supposed to offer help or not. Are we doing the wrong thing if we speak to the person?"

"I always appreciate it when someone asks if I'd like some help, especially if I'm crossing a busy street," replied Gretchen. "If a blind person says yes, then offer your left arm—don't try to take hold of the person. And when you get to the other side of the street, don't just instantly say, 'Well, good-bye,' and take off—make sure he or she is headed in the right direction and is in control before you leave. I always say, 'I'm okay now, thanks,' when I'm ready to be on my own again. It takes only a few more seconds, and we appreciate that kind of considerate help."

Some people offer to help blind people, others ignore them. I recall a time a couple of years ago when I helped a blind person with a guide dog get a taxi. There was this guy, gamely trying to flag one down, with his dog waiting patiently at his side. The cabs just weren't stopping for him. I waited till an empty cab stopped for a light, then I ran and opened the back door.

"I have a fare for you," I said to the disgruntled cabbie, and led the man and his dog into the taxi, wished them well, and closed the door.

That taxi driver was a real jerk, but he's not alone in representing a widespread attitude toward handicapped people: It might be called handicapist, like racist or sexist—a view that handicapped people are different, inferior to the rest of us, and therefore should be ignored as though they don't exist. Blind people are changing handicapist attitudes themselves as more of them enter the mainstream of society. Gretchen may be blind, but she certainly doesn't let you pity her or think of her as handicapped.

I never thought much about disabled people before I got into this field, but now I'm pretty aware. Training dogs for the blind and teaching the blind people how to work with their dogs, well, if you want a job with a lot of what they call ego gratification, this is it.

4

THERAPISTS
IN THE BARN

Steve tells
the story of his year
as a counselor
at a farm school
for troubled children.

I was relaxing in the staff lounge on my lunch hour when a
boy burst in, out of breath.

"Steve, Steve, come quick," he gasped. "Tim needs
some help in Cottage Nine."

I got over there on the double. Cottage Nine houses
about a dozen boys aged ten to thirteen with two resident
counselors, Tim and another fellow. I knew if something had
happened in his cottage, Tim might have his hands full.

Serious fights among the kids here are rather rare, but
there was mayhem in Cottage Nine. Tim was peeling boys
off a screaming youngster on the floor. I saw it was a kid
named Kevin. As fast as Tim would get an armful of struggling boys, other kids would tackle Kevin and start punch-
ing him. Everybody was yelling, of course.

I grabbed Kevin and dragged him out the front door

51

fast, closing it behind us and holding it shut. The frightened boy checked himself out to see how badly he was hurt. I could hear Tim trying to calm the bedlam inside.

"They all jumped me," Kevin said. His nose was bleeding a little.

"Why?" I asked him. "What did you do?"

The boy looked at me defensively. "It's because of the duck."

It looked as if we might have to put a kid in protective custody because of a duck.

Apple Tree Farm is temporary home to ninety-odd of the most troubled, confused, and sometimes off-the-wall kids from the city. Set in rolling countryside, it was started by the state several years ago for use as a rehabilitation residence and school for disturbed children who have come to the attention of the police, courts, or social-service agencies. These youngsters need help badly. Some of them have come from horrendous family life with abusive, drunk, addicted, or violent parents. Some have no families at all.

Ranging in age from about five to fifteen, the boys and girls spend two years here in the care of a large professional staff who are determined to heal them and help them become functioning and self-controlled human beings. Our work with these difficult children sometimes seems hopeless and futile; many of them are very damaged when they arrive. "Future Murderers of America," one of the counselors once called them jokingly when we were among ourselves in the staff room after a particularly bad day. But inside, we are all true believers. We are really convinced that these kids can be helped—and apparently many are. We often see it happen.

52

The children receive special schooling, psychiatric treatment, and supervision from a hard-working bunch of educators, therapists, and counselors. I am an intern here, serving as a junior counselor while earning credit for my degree in social work.

And one of the most important parts of my job is supervising the kids with the farm animals. I was born and raised on a farm, so my background comes in handy. There are one hundred fifty cows, horses, pigs, sheep, goats, rabbits, chickens, ducks, and geese here. These animals are not pets—Apple Tree is a working farm. There are a few good-natured dogs and cats who qualify as pets. And taking care of all the animals, under supervision, is part of the youngsters' therapy.

Why does taking care of farm animals help these kids? We're not sure exactly why, we just know it does, perhaps for a combination of reasons. For one thing, these are city youngsters who have lived where starving and abandoned animals roam the streets, where pet animals are often neglected or actively abused. The idea that animals are valued and treated well is amazing to them. Learning to respect these animals' rights and needs, we think, is a step in the direction of learning to respect the rights and needs of other human beings.

Also, for children who usually lack verbal skills, who may not speak well or express themselves clearly in words, the animals are a relief. Animals not only express themselves mostly in body language, but they respond to the body language of people far more than to words. With the animals, the children don't have to know how to use verbal language well.

The animals also offer an opportunity for play to kids

Farm animals can help bring out feelings of affection and responsibility in young people. This boy and a donkey are sharing a moment of wordless communication.

who may never have had a chance to be childlike. A youngster who has had to be worldly and streetwise almost as soon as he or she could walk has missed an important part of childhood. Many animals invite and respond to simple play.

And the animals themselves, by their very dependency and their noncritical acceptance of the youngsters, seem to make the kids who care for them feel important. Many of the animals give affection. They also offer a chance for real pride—the kids enter animals in 4-H competitions and the county fair and often win prizes.

Take the case of Josephine. When she came here to Apple Tree about six months ago, she was sullen, withdrawn, and vastly overweight. She refused to have any part of the work with the animals. Somehow, a riding instructor named Shelley got her down to the stable one day, and when Josephine's eye fell on a beautiful black mare, something must have awakened in her. She began coming down to the stable, helping feed and brush the horses and muck out the stalls, watching the classes getting ready to ride, that sort of thing. One day she asked Shelley to teach her to ride. It was the first time Josephine had wanted to learn anything in the entire school program. Not only has she now learned to ride—she has slimmed down to normal weight. She's practicing to ride the mare in the local horse show. We all hope Josephine will win a ribbon; she has never won anything in her life and she is working hard.

Some of the riding classes are a sight to see. I'll never forget the first day I went down to the stable with the group of eleven- and twelve-year-old boys assigned to my cottage. Into the tack room we went, and immediately the

55

kids began to quarrel loudly over who would ride which horse. They all wanted Nugget or Odyssey; nobody wanted Sam. "He threw me last time," one boy complained bitterly.

Shelley and the other riding teacher, Carol, listened to everyone and then assigned horses. There was a lot of grumbling, but the kids hoisted saddles from the racks and trooped into the stalls, jostling and pushing one another. Their language was really gross, and they swaggered among the horse stalls so full of macho that I began to worry they might actually abuse the horses. They threatened the animals—"You bite me, Sam, and I'll cut up your eyes!" But these horses, while gentle enough to be ridden, will defend themselves by nipping, kicking, or bucking if mistreated, and the boys apparently know it, so their behavior was pure bravado.

Shelley and Carol worked like fiends, giving encouragement and helping the boys saddle and bridle the horses, and somehow keeping control of the scene. Their patience was monumental.

One by one the kids led the horses out into the yard. Each boy managed to get himself on his horse one way or another, and they walked down to the riding ring. They were the oddest assortment of riders I'd ever laid eyes on, in terms of looks and style. And yet, once they were all milling around in the ring, peace settled over the group like sunshine. The change in the boys was amazing. Minutes before, they had been cursing, quarreling, and bragging. Now they were actually smiling, cooperating, and making real efforts to improve their riding.

"Shelley, look at me, am I doing this right?" called one.

"Carol, Carol, how do I get him to trot?" another hollered.

In a short while, they left the ring single file for a trail ride through the woods, Shelley in the lead and Carol bringing up the rear.

Horseback riding is considered extremely important for the children. The mastery of riding gives them confidence in themselves, and learning to control such a large animal as a horse helps teach them self-control. And for city kids who have seen only privileged people riding in the parks, being able to ride must raise their self-esteem.

Because it's a working farm, Apple Tree provides its own eggs and much of its own meat. That means some of the animals the kids have raised and cared for are slaughtered, butchered, cooked, and served. The staff is very firm about this, and therefore doesn't encourage making pets of animals that are possibly going to be killed. It's difficult for some of the children when this happens, and certain ones stop eating meat. I'm a vegetarian myself, and so are a few of the other counselors and teachers.

The reason I don't eat meat is that most of the animals from which meat comes nowadays are raised in appalling conditions on factory farms. The animals spend their entire short lives in huge buildings, kept in darkness in tight confinement, unable to move or turn around. At least, the animals at Apple Tree have had normal, comfortable lives. And they are not crammed into trucks and driven across the country to slaughterhouses, but are killed humanely here by a professional (out of sight of the youngsters and staff).

Nevertheless, whenever one of the animals is slaughtered, some of the youngsters are outraged and threaten to "liberate" the other animals to save them. We have some heavy discussions about the whole subject of raising animals for meat. Most of the children and staff have no hesitation about eating meat, and even most of the kids who complain when animals from our farm are killed fall back into the habit. After all, the youngsters come to the tables hungry and meat is served to them, so you can hardly blame them. But there are a few holdouts. One kid in my group always sits next to me at the table for moral support as a vegetarian.

This concern for the animals is truly a breakthrough for many of these youngsters. However, sometimes it creates situations such as the attack on the boy in Tim's cottage.

Some of the boys had been down at the pond one evening, throwing stones in the water, when a few of the ducks swam in sight. Kevin threw a rock at one of the ducks—and hit it. The bird floundered in the water, the kids yelled, and Tim rushed over and waded into the pond and picked up the duck. Its wing was hurt. Tim made it a bed on some hay in a box and put it in a quiet place for the night, planning to take it to the veterinarian the next morning if it wasn't getting better. During the night the duck died. Word spread through the school, and Kevin was in big trouble with the other kids. After the fight in the cottage, we made sure Kevin was never alone for several days until the worst of the anger against him subsided.

Some of the children arrive here so emotionally disturbed they seem to have no feelings at all. Others are trigger-quick to show rage when anything bothers them. So I was really encouraged when a particularly wild-eyed, difficult little kid in my group formed a healthy attachment for one of the cows. This boy was named Raoul, a handsome, black-eyed twelve-year-old child who was small for his age but strong and energetic. Sophie, the cow, was pregnant, and Raoul began to take an interest in her well-being and comfort. He brushed her, gave her extra food, cleaned out her stall several times a day, and made sure she always had plenty of clean hay. Raoul had never before, to my knowledge, showed such serious concern for any living thing— person or animal.

"Are you going home for Easter, Raoul?" one of the counselors asked him one day as Raoul was carefully shoveling manure out of Sophie's stall. Raoul had a large, disorganized family somewhere in the city. He was mildly fond of his younger brothers, indifferent to his mother, and afraid of his father, I had learned.

Raoul ran his hand lovingly over the cow's bulging side. "Leave her?" he said. "No way!"

The kid actually refused to go home when the long Easter weekend came. Apple Tree Farm was fairly quiet because quite a few children who had homes to go to had left for the holiday.

One evening I checked on Sophie after dinner, and she looked to me as though the blessed event was going to take place any minute. Because of my familiarity with animals, I offered to sleep in the barn, just in case Sophie needed any help.

"I'd appreciate it if you would, Steve," said the farm manager. "Call me if there's any problem, though."

I got permission from the senior counselor in my cottage, got my sleeping bag, sacked out near Sophie's stall, and went to sleep. Things began to happen about three in the morning. At one point, I thought I might have to call the farm manager, but Sophie did okay by herself with only a little help from me. Dawn was just breaking when her

59

baby emerged safe and sound. As I was looking at the cute little wet calf lying in the straw, I had an idea. I raced up to the cottage.

"Raoul," I whispered, shaking the sleeping boy. "I've got something to show you." Raoul opened his eyes, and when he saw it was me, he jumped up like a shot.

Raoul beat me to the barn. When I walked in, he was standing absolutely still in Sophie's stall, watching her clean the calf, who was just struggling to its feet.

I glanced at Raoul, expecting to see him all smiles. But Raoul was crying. Big tears were rolling silently down the face of this tough little kid, this so-called juvenile delinquent who, according to his social worker's report, had terrorized his neighborhood. Sophie the cow had done what none of us with all our professional training and skills had been able to accomplish: She had helped Raoul feel something beautiful.

An interesting situation came up recently with a group of the younger kids, boys and girls about six or seven years old. The pond on the farm property has lots of frogs and little turtles, and sometimes the children catch them. One Sunday a group decided it would be fun to have a frog-jumping contest and turtle race.

Kids had fanned out around the pond and in the marshes, looking for frogs and turtles to capture. I rounded them up, saying I wanted to talk to them. There was a lot of grumbling and resistance, and a few refused to come.

"Frogs and turtles may not be as smart as people or dogs, but they have feelings just the same," I began. "When you capture them, you scare them. A frog or turtle will instinctively want to get away and hide. To such a small animal, you must look like some kind of giant monster. These

Learning to care for and respect animals can help a young child develop empathy and other civilizing qualities. This little boy handles a kitten with tenderness and self-control.

animals just want to be left alone and to live their lives in peace."

The kids squirmed and mumbled among themselves. "But we don't really hurt them," one girl ventured.

"You may not mean to, but in handling them, you could accidentally hurt them," I said. "If you go ahead with this contest, some of these creatures will die. You will unintentionally break their legs, or bruise them, or they may just die from fright."

Silence from my audience.

"Look at it this way," I continued. "Wild creatures are greatly harmed by being handled or forced to do things that are unnatural for them. To make the frogs and turtles jump and crawl the way you want them to, you'll have to tease and hassle and scare them. Don't you think they have a right to be left alone in the pond? How would you feel if you were them?"

I got through to some of the children. They wandered off to occupy themselves with some other kind of play. A few went back to catching frogs and turtles—but the race never came off.

I know it's asking a lot of children who have been mistreated to accept the idea of the rights of others—in their past lives, nobody respected *their* rights. However, most of them seem to be able to accept the idea of the rights of animals fairly soon, even though it's a new concept to them. Compassion for animals seems to come more easily than empathy for their own species, perhaps because animals have never harmed them or perhaps because they can identify more easily with animals as fellow underdogs. We hope it's a starting point.

I've heard of an interesting project involving animals and mentally or emotionally disturbed adults in a psychiatric hospital/prison in Lima, Ohio. At this hospital are people who have been judged insane in court proceedings or who have cracked up while serving time in state penitentiaries.

One day about seven years ago, a psychiatric social worker there named David Lee noticed that some of the men on his ward were trying to save a wounded bird they had found on the grounds. They were so interested and concerned about the animal that it gave Lee an idea. He got permission and some money from the hospital superintendent to buy a tankful of fish and two parakeets. The patients became very nurturing toward the animals, and even just those few pets improved the atmosphere and morale in the ward.

So then Lee decided that patients who wanted pets could earn the right to have their own, if they were well enough and responsible enough to take good care of them. The pets were all of a type that could live comfortably in cages in the patients' rooms—birds, hamsters, gerbils, and guinea pigs.

Well, the project took off. Men and women who had never before cared about animals became aware and concerned about the needs of other creatures. And most importantly, the pets helped the communication between the patients and the therapists who were trying to treat them. The patients became more trusting of the therapists, and everybody talked about the animals. Also, many very sick patients were motivated to try to get well enough so they could have pets. Depressed patients, many of whom had attempted suicide, seemed to have found a reason to live.

Now there are several hundred animals living at the hospital prison. All were donated from the outside, except those that were born in the hospital. The owners earn the money to feed them, and a local veterinarian donates his services when needed.

They've even acquired some animals that live outdoors in the courtyard—two deer, some rabbits, and a goose. One of the deer came from a zoo where it had been stoned by vandals. It was not recovering well, so the zoo director gave it to David Lee. The patients nursed the little animal back to health, and today it is living in the courtyard with the other outdoor animals.

It's interesting that these patients, some of whom have committed violent crimes in the past, become so tender toward the animals, and so proud of their pets. Nobody has ever harmed an animal there. The effect of Lee's "pet therapy" program has been dramatic.

"Animals will be your friends," a patient once said. "They don't care where you're coming from, they'll like you anyway."

Here at Apple Tree Farm, with youngsters who have been mistreated and who have a high potential for mistreating others—human beings *and* animals—the humane stewardship of animals is our first goal. Perhaps this can be the foundation for a greater appreciation of animals later on. I was effective that Sunday in impressing some of the younger children with the concept that frogs and turtles have rights. Whether this will stay with them, I don't know. I hope so.

There's no doubt that they recognize the feelings of our farm animals. We had one very funny episode with one of the youngest children this spring. This little boy, Tony, loved the farm animals, and we had to keep a close eye on him in the barn. He was totally unafraid, would go right into the stalls and pens. We were concerned that he'd be stepped on or kicked, and we supervised him carefully.

When another pregnant cow went into labor one morning, about a week after Sophie's calf was born, Tony was right there, hanging on the gate to her stall. So were all the children who were free from classes or chores—they rushed to the barn to watch. To these city-bred kids, a cow giving birth was surely an unusual sight.

This cow's labor was difficult, and suddenly we knew something was very wrong. The calf was not in the right position in the birth canal. A calf normally comes diving out of the mother head and forelegs first. But in this case, we could see all four tiny hoofs sticking out. The cow was straining and pushing, but obviously there was no way she could push the calf out in that position. She and the calf were both in danger, so the vet was called and asked to hurry over. He arrived, rolled up his sleeves, put on gloves, reached in, and carefully turned the calf into the right position. Then it was delivered safely. The youngsters watched in goggle-eyed silence, especially Tony. They were really impressed with this lesson in bovine obstetrics.

The children were oh-ing and ah-ing over the calf, and the vet was getting ready to leave, when suddenly Tony burst out, scowling and waving his fists.

"Who stuck that little calf up inside the cow that way?" he roared indignantly. "That's mean! I bet it was one of those boys in Cottage Nine. If I find out who did it, I'm going to tell on him and get him put in the slammer!"

65

Funny how we all had just assumed these children know everything from having grown up in the streets. Apple Tree doesn't have a "slammer," of course, but Tony certainly wanted the mean person to be punished. Shelley hugged him and led him off by the hand, and perhaps she gave him a simple lecture on the facts of animal life so he would understand better what he had witnessed.

Apple Tree Farm's animals serve as fine teaching aids. I like to think they also show Kevin, Josephine, Raoul, Tony, and the others something of the connection of human beings to other forms of life, so they can understand how we are all related and dependent on each other in this world.

5

PETS AND THE HEALING PROCESS

Karen describes her work
as a student teacher
at a school for preschoolers
with emotional problems.

When I was a teenager and did a lot of baby-sitting, I thought I had seen every type of little kid. And in the months since I've been a student teacher in this nursery school, I've worked with all kinds. But I had never encountered a child as withdrawn as three-year-old Chris. He was totally unresponsive. He stiffened like a board when we picked him up, showed no reaction to hugs and kisses, and pulled away when the other children tried to play with him. He never smiled, laughed, frowned, or cried. He rarely spoke.

The other teachers told me that when he was a baby, Chris had been abandoned by his mother to the care of a crazy aunt who had kept him in his crib most of the time, drugged so he wouldn't cry. When the child protection authorities discovered him and removed him to a foster home,

67

he learned to walk and became healthy enough physically, but not emotionally.

At school, he moved aimlessly around the room, picking up toys and putting them down, always in motion but with no goal in his activity. Every day, we tried to interest him in the other children or in the animals, but Chris had locked a door on himself.

One day I tried to get him to pet Bozo. I put my arm around him and called the dog. "Here, Bozo," I said, and the animal came at once, wagging his tail.

"Look, Chris, isn't Bozo a nice dog?" I asked, taking the child's hand and running it over Bozo's soft fur. No reaction from Chris. As far as the child was concerned, the dog did not exist.

If you looked around the playroom of the school where I work, you'd wonder what is going on. Bozo lies on his belly on the floor while a little boy sits astride him and another child hugs the dog around the neck. A small boy and girl walk by pushing a doll carriage in which lies an uncomplaining cat dressed in doll clothes—paws through the sleeves, ears tucked under a bonnet. A group of children are building a complicated maze with blocks on the floor for two hamsters who scamper through. Three other kids are playing with a doll house—suddenly a rabbit pokes its head out one of the windows. Everywhere, everywhere, children and animals.

Many nursery schools, of course, have a few pets—gerbils or guinea pigs, maybe a tankful of fish. We have dogs, cats, rabbits, hamsters, parakeets, and a dove, and the animals mingle freely with the children every day. We call them our teaching assistants. In fact, sometimes Bozo can handle a kid better than I can.

The children at this school all suffer from some kind of emotional problem. Some of them have been neglected or abused and, like Chris, live in foster homes. We think the pets are important aides in the education and treatment of the children. According to the therapists who work with these children, the kids talk to them more freely about their problems when the animals are present. And the children also talk to the pets, and the therapists pick up a lot by listening to what they say and observing the way they play with the animals. The use of the animals in this school is part of a rising trend in the therapy of disturbed children. It even has a name: "Pet-Facilitated Therapy."

"A child who is anxious and withdrawn, who won't talk to the psychiatrist or psychologist who is trying to help him, will often open up to a dog or cat in the presence of the therapist," one of the teachers told me. "The child feels safe talking to the dog or cat. He knows the animal won't criticize him or reject him, no matter what secrets he tells it. The child feels freer to tell the animal what's bothering him than the therapist."

I once read a book about a famous dog named Skeezer who was the first known animal to work as a "therapist" for children. This was a true story. Skeezer, a big, sweet-tempered mixed breed like Bozo, lived at a children's mental hospital in Michigan. She made being in the hospital a lot more bearable for many frightened, disturbed children. She was also apparently a valuable aide to the doctors and nurses who were treating and caring for the young patients.

In psychiatric hospitals where pets live or visit, the animals serve as adjuncts to the other forms of therapy provided. But it is interesting that time and again, when other forms of therapy have not worked, the animals serve to

69

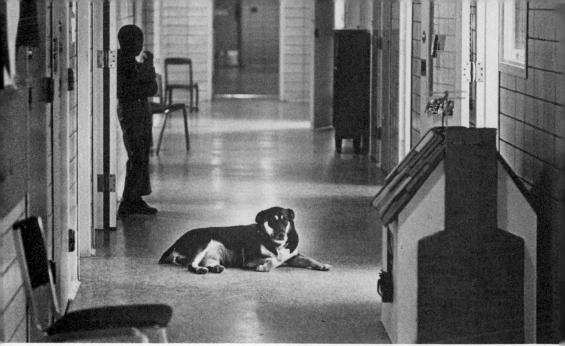

Skeezer was a respected therapeutic aide and made the hospital a more humane, less scary place for the young patients. She had her own house in the hospital corridor and was never too busy to listen to a child's secrets.

break the ice and facilitate the other. Many cases have been reported of patients who were helped by medication and other treatment only after pets were introduced into the regimen.

I've always loved animals and children, and it suits me just fine to be teaching here for credit in my special-education courses in college. I had a hard time deciding to become a teacher instead of a veterinarian like my sister Lisa. When Lisa was in veterinary school at the University of Pennsylvania, she told me about a long-term study going on there of the relationship between people and their pets.

The bond between people and pets—the "human/ companion animal bond," as scientists and scholars call it— has more or less been taken for granted and never given much serious thought—until researchers made a discovery that suggests that pets are not only our good companions but beneficial to our health!

Not long ago, some researchers made a thorough study of a large group of patients who were suffering from chronic heart disease. A year later, the researchers interviewed those patients who were still alive. They compared all the factors and variables, and found that most of the people who were still living had pets. Most of the patients who had died over the year had no pets.

"The study raises some good questions, Karen," said Lisa. "Maybe the animals gave their owners something to live for. Or perhaps just having to get up and walk around and take care of their pets gave the pet owners the exercise they needed. Or—it's possible that the type of person who has an animal in the first place might also have some built-in personality factor that contributes to survival. And possibly the simple, undemanding love a pet gives is beneficial in ways we are just beginning to discover. Anyway, a group at my school is investigating the true nature of this bond between people and their companion animals."

Some interesting findings have begun to turn up. One research assistant, for example, studied people waiting with their pets in the reception area of a veterinary clinic. The student used her own dog as a cover and pretended to be doing homework while she made notes, so she wouldn't appear to be watching the clients with their animals.

What she observed was that all the men and women stroked and fondled their animals to calm them, because the pets are, of course, nervous at the veterinarian's. But it also seemed that the clients were trying to calm their own nervousness as well. You know how babies like to fondle a toy or a favorite blanket for reassurance when they're going to sleep? It's possible the clients pet their animals for the same reassurance.

"From past studies, it's known that being stroked has a calming effect on dogs. It lowers their blood pressure, slows their breathing," Lisa explained. "But nobody had ever studied the effect on the person doing the petting. We've established that petting an animal has a soothing effect on the person and lowers his blood pressure as well."

One thing the researchers know for certain—talking to your pet does not produce the physical reaction in you that talking to another person does. They have measured the blood pressure of people talking to each other and learned that no matter what the conversation is about, there's a slight rise in their blood pressure. But when the researchers took the blood pressure of people talking to their pets, there was no rise in blood pressure. The knowledge that pets are uncritical, nonjudgmental listeners certainly bears out what the therapists at our nursery school have found in observing the children talking freely to the animals.

I heard about a resident hospital for the retarded in Louisiana where a compassionate and wise psychologist introduced dogs to provide affection, interest, and stimulation for the residents. The animals live on the grounds of the hospital and are cared for by several of the more intelligent and able patients, who love them and see to it that they are treated well by everybody.

"We brought in young, very friendly, healthy dogs who would be happy living in this setting with so many people around," the psychologist told us at a lecture. "We slowly and carefully taught certain patients how to care for and be responsible for the dogs, under supervision by the staff. Almost all the patients pet and hug the dogs, play with them, or watch them. The dogs are the most important event in this place.

"And it's amazing to see how the dogs react to the people," he went on. "With the less retarded, the animals are bouncy and playful. They are not put off by any disability. Some of the patients can't help grimacing, drooling, speaking strangely, or making jerky motions. The dogs don't mind a bit—they don't even seem to notice. They are not critical of intellectual inferiority.

"But when the dogs are with the more profoundly retarded patients, they become very quiet and gentle. They don't jump up on these people, play-bite, or frisk around, but wag their tails and lick the people's hands and faces and stand very still while they're being petted. The dogs seem to know these people can't play with them the way the others do, but need special treatment."

An early work with animals and the mentally handicapped was conducted in the 1970s by two scientists, Samuel Corson and his wife, Elizabeth. They introduced dogs into a nursing home for retarded and intellectually disabled people in Ohio, and documented the dramatic improvement in many of the patients. We've heard of Seeing Eye dogs and hearing ear dogs—well, Dr. Corson said animals that help psychiatric patients this way should be called feeling heart dogs.

In fact, Chris showed his first sign of affectionate behavior at school with Bozo. We found Chris one day curled up on the floor in a corner, snuggled against Bozo, one arm over the dog. Chris still didn't speak, but Bozo had accomplished a breakthrough.

Bozo had been bailed out of the local pound by one of the teachers, who had him checked over by a veterinarian and then took him home. When she saw what a sweet temperament he had, she began bringing him to the school. He's been here ever since. Bozo is no beauty—in fact, he is a bit homely, a big sturdy animal—but he has a thick, soft, black and white coat that children love to pet. He is so good-natured, the kids can lie on him and roll around on the floor with him, and he'll just grin, wag his tail, and lick them. In fact, when they're tired of playing with him, he'll tag them around the playroom or yard anyway.

We also have a mellow, female mixed-breed spaniel named Dandy, and a little dachshund, Fritzie. The four cats were adopted as kittens, so they have grown up accustomed to lots of handling and cuddling. Fluffy is the one who'll tol-

74

erate being dressed up in doll clothes. I think these animals understand a lot.

It is interesting and somewhat surprising to the other teachers and me that there have been only rare occasions when we have had to interfere to keep an animal from being hurt or pestered by these children. We watch closely. In any rehabilitation work using animals, abuse should always be anticipated so it can be avoided or stopped. Mistreatment of the pets might be a problem in some instances. We have not encountered it.

Just recently, we had another hopeful signal from Chris. We had gathered the children into a cluster for quiet play after lunch. Bozo, Dandy, and Fritzie were sprawled on the floor near us, taking naps. Chris was off by himself as usual, wandering aimlessly around the room.

Suddenly one of the young cats spotted a fly at the window. Whiskers twitching, it jumped down from the desk where it had been sunning itself and began to stalk the insect.

"Look at Chris," a teacher whispered to me.

Sure enough, Chris was watching the cat. He was actually looking, focusing on the animal.

Wiggling its rear end, the cat crouched, then sprang at the window. It bounced off the window and landed gracefully on its feet, but in the middle of a big flowerpot full of plants. The pot tipped over, and the astonished young cat made a scrambling somersault in midair on its way to the floor with the flowerpot.

We heard a peal of laughter. It was Chris. Welcome to the world, Chris, I thought.

When I graduate, I will continue to teach mentally or

75

emotionally handicapped children, and I hope I'll always work in a school where animals are a big part of the scene. I think people are only beginning to understand the nature of our bond with our pet animals. The new research certainly suggests that they are beneficial to our health.

Who knows—some day very soon it may be commonplace for a doctor treating a lonely, depressed, or withdrawn person to write out a prescription that says: "One dog or cat, to be stroked for fifteen minutes every four hours, daily."

6

EARS WITH ORANGE COLLARS

Vicky, who tells
her story here,
is a student trainer
of hearing dogs
for the deaf.

The minute we walked into the animal shelter, all the dogs began to bark. Some flung themselves at the doors of their cages. Most stood on their hind legs pleading desperately to be taken out. "Me! Me! Choose me!" each one seemed to be saying. "Get me out of here!"

Even more pathetic were the few who had given up hope and just lay at the backs of their cages with their faces to the wall.

I walked down the aisle between the cages, looking over the dogs. I was searching for a very special dog to adopt, and I hoped I'd recognize it when I saw it. The dog I wanted had to be young, small-to-medium sized, healthy, smart, friendly, and willing to learn.

The trouble was, I wanted to take *all* the dogs. I knew that some would be adopted and go to good homes; a few would probably be adopted to terrible homes where they

77

would not be treated well. But most of them would be euthanized or "put to sleep" here at the animal shelter because nobody adopted them—otherwise the shelter would become overcrowded. There are always many more dogs brought in than there are homes available for them.

I stood helplessly amid the confusion and noise. Just then Mr. Caprano came along. He was my teacher, whom I had come to the shelter with. "What's the matter, Vicky?" he asked. "None of these dogs seem right to you?"

"It's hard to decide," I mumbled, feeling rather foolish. I was supposed to be businesslike and professional; instead, I was afraid I appeared bogged down in sentimentality. But Mr. Caprano seemed to understand.

"I want to take all of them myself," he said. "But come and look at this little mutt I found over here and tell me what you think of her."

He led me to a cage where a fluffy little brown and white dog gazed out at us. She was about the size of a large sheltie or very small collie. She was alert and friendly, and she wasn't barking or acting quite as hysterical as most of the others. She looked as if she fully expected something good to happen.

"Let's have a look at this dog here, please," Mr. Caprano requested of the shelter manager, who took the dog out of the cage for us. The animal put her paws up on me and licked my hand. Then she went and looked into Mr. Caprano's face, wagging her tail hard. This little pooch really knows how to sell herself, I thought. Who could resist her?

"She was noticed sitting on the porch of an empty summer cottage near here," said the manager. "After three days, someone in the neighborhood brought her to us. Obviously, her people had simply gone off and left her.

"She was very hungry and thirsty, but otherwise in pretty good shape," the woman continued. "We think she's about a year or eighteen months old. She's very good-natured."

I could see Mr. Caprano was sold on her, and I was too. Just then I noticed the dog in the cage next to hers. He was standing as close as he could get to the front of his cage, watching us intently. He was a short-haired, black male dog, somewhat taller than the brown female but lightly built. He had one ear that stuck up and one that hung down, giving him a cute, quizzical look. There was a very gentle expression in his eyes.

"Who's this one?" I asked the shelter manager.

"Oh, he was found with her," she said. "We think they had belonged to the same family, who abandoned them both. The two dogs were obviously friends. He's a little older, maybe a year or two. He's a very nice dog, but you wouldn't want him—he's lame."

Then I noticed that one of the dog's hind legs was missing just below the joint. "Probably got caught in a trap," the manager commented. "He limps, of course, but I must say he gets around as well as any dog. It doesn't seem to bother him that his leg is missing."

I looked at Mr. Caprano and he looked at me. How could we take her and leave him here?

Of course we took them both.

While Mr. Caprano filled out the forms and made a donation to the shelter, I walked the two dogs out to the car. It was true, the black dog seemed hardly handicapped at all. They were overjoyed with their freedom and piled into the backseat without any hesitation at all.

When we got to the college, a group of students came

79

out to see the new dogs. The black male stood waving his tail, accepting all the petting with dignity.

The little brown and white mutt flopped on her back and grinned as Peter, one of my classmates, knelt down and tickled her.

"You're about to start a very interesting life," said Peter to the little dog. "If you play your cards right, you're going to become a hearing ear dog."

When I came to Collins Junior College, I knew I wanted to go into some work involving animals—maybe work for a veterinarian, an animal hospital, a breeder, or a zoo. Then I discovered that Collins has a hearing dog program, which Mr. Caprano directs. I'd seen guide dogs for the blind, of course, but I had never heard of hearing dogs. Students who major in kennel management here can enroll in the course to learn to train these dogs and teach the deaf persons who are selected to receive dogs.

It is exactly the right field for me. I love working with the dogs. Also, I have a hunch that the hearing ear dog idea is catching on and there will be an increasing number of agencies looking for qualified persons to teach, so I think I'll find plenty of job opportunities.

Some people are attracted to this work, as I was, because they like to work with dogs. Some have professional experience with the deaf. I intend to go on eventually and get my college degree in education with an emphasis on teaching the deaf.

There are some fifteen million totally deaf or hard-of-hearing people in the United States. In many ways, their disability is worse than blindness. For one thing, a blind person has a 360° knowledge of what's going on around him, because he can hear from all sides. He can hear what's happening in the next room, across the street, over the telephone. A deaf person can only be aware of what's in front of his eyes.

A deaf person's disability is hidden, and strangers don't realize that he can't hear a shout, a siren, a question, a warning. Deaf people have even been shot by the police who ordered them to stop and thought they were ignoring the order. I never really realized how severe the problems of the deaf are until I got into training dogs for them.

Some people working to improve the education of the deaf believe that they are better served if they are taught to communicate by means of sign language rather than lip reading. These educators say that signing is the natural language of the deaf. There are several sign languages. One is word spelling. Another is American Sign Language, or Ameslan as it is called, which has signs for whole concepts or ideas. (The Ameslan sign for *dog*, for instance, is a slap on the thigh followed by a quick snap of the fingers.) The various sign languages enable deaf people to communicate with one another and with others who have learned those languages.

Other educators of the deaf believe that a person should learn and use all methods. They say deaf people are better off if they are equipped with every possible tool to communicate, including the ability to lip-read, speak, and use any of the sign languages.

Here at Collins Junior College, we in the hearing ear dog program learn Ameslan. The deaf people who arrive here for dogs come from a variety of backgrounds. Some can speak and lip-read, some can hear a little with hearing aids,

81

others know different sign languages, but we manage to figure out ways to communicate with all of them once we have the basic knowledge of Ameslan.

For the convenience of deaf people who speak, we give the hearing ear dogs names that they can pronounce easily—Buffy, Polly, Rover, Bobo, Meg—nothing fancy like Aristotle or Olivia. We named the new little brown and white dog Meg and the black male dog Buck.

The morning after Mr. Caprano and I brought Meg and Buck to the college, I took them to be tested by our senior trainer. She gave them some simple obedience lessons to see how they responded. This teacher is always wonderfully kind and patient, but firm—dogs pick up instantly that she expects them to obey, and they rarely fool around. Meg threw herself into what she must have considered a great game, being asked to come when called and made to sit, even lie down. Our trainer thought she would do well—she was eager to learn.

Buck, on the other hand, showed that he must have had some obedience training before. He even responded immediately to the hand signal to lie down. Or else, because Meg was tested first, Buck caught on from watching her—is that possible? I wondered.

Anyway, both dogs looked promising, and their training in obedience was begun at once. Peter and I were assigned to give Meg and Buck their obedience training, and then we would continue their advanced work under supervision.

Because many people who were born deaf have trouble speaking clearly, the dogs are taught hand signals rather than verbal commands. They learn to sit, stay, come, heel,

lie down, fetch, and the like. We give the dogs twenty minutes of work morning and afternoon—that's about as long as
they can concentrate and pay attention without getting
tired. Praise and petting when they respond correctly to a
command; repeated work when they don't catch on; but
never punishment—that's the system we use. The senior
trainer says timing is highly important in teaching dogs.
You have to be on your toes to pick up signals from the animal and correct it immediately when it makes a mistake,
and reward it instantly when it catches on and carries out a
command correctly.

Meg was so eager to be petted and praised that she
really put a lot of effort into learning. Buck, however, continued to surprise us. He worked seriously, mastering the
obedience training so fast that we didn't waste time repeat-
ing lessons once it was obvious he knew them. He was
ready for the second stage of training before Meg was.

Hearing ear dogs are taught to alert their deaf owners
to significant household sounds—the alarm clock, doorbell,
smoke alarm, teakettle, oven timer. Some dogs are taught
telephone response. They are given to deaf people who can
hear on the telephone or on an amplified phone, as some
can, or who have special TTY (teletype) telephones. And a
dog that's being trained for a deaf couple with an infant will
learn to fetch the mother or father when the baby cries.
(We even heard of one smart dog that fetched her deaf
owner to her ten-year-old daughter's room late one night.
The woman discovered her child sitting under her desk
reading by flashlight. The dog knew it was past bedtime for
the girl!)

The dogs alert their owners very insistently—they

Above: An alert and lively dog is responding to a knock at the door. It will learn to bring its deaf owner to the door whenever someone knocks or rings the bell.

Below: A trainer is teaching a hearing ear dog to respond to the phone, while the other trainer plays the role of a deaf person, whom the dog will learn to "tell" when the phone rings.

don't give up until they get results. At an alarm-clock signal, for instance, the dog jumps on its owner's bed and paws at the covers or licks the person's face, or both, until the person is out of bed and standing up. The response to the doorbell is to run back and forth between the owner and the door. But the response to the smoke alarm is to rouse the person and lead him or her out of the house.

Outdoors, the dogs learn to "tell" their owners of automobile horns, sirens, shouts, or other warning sounds by jumping up on them or stepping in front of them. Dogs that will be given to deaf people who drive are taught to put a paw on their owner's shoulder when a siren is blowing or another car is honking.

The idea of dogs for the deaf originated at the Minnesota SPCA, where a professional dog trainer named Agnes McGrath trained several dogs in a pilot project in 1973. A few years later, Mrs. McGrath developed and directed the first systematic hearing ear dog training program at the American Humane Association in Denver. She now runs her own training center in Henderson, Colorado. Altogether, there are maybe eight or ten agencies around the country that train such dogs, but more are starting. Although there's a great demand for the dogs, and most places have waiting lists, the growth of programs for training them has been hampered by lack of money. It costs an estimated two thousand dollars or more to train a dog. Some agencies are funded through contributions, however, so they can charge the deaf recipients less, or nothing.

Pete and I worked with Meg and Buck one at a time, in a training cottage that was furnished to resemble a real home and was equipped with all the necessary bells on a control panel. We started with Buck and the alarm clock. One of us would lie down on the cot in the bedroom of the cottage and set the alarm clock while the other waited in the next room with Buck on the leash. Let's say it was Pete's turn to lie on the cot and set the alarm clock. When it rang, I jumped up and led Buck into the bedroom, saying, "*Sound*, Buck! Tell Pete." At first Buck seemed reluctant to jump up on the cot, but I encouraged him, and then we both petted and praised him.

Soon it was possible to let Buck off the leash, and when the alarm clock went off he would gallop into the bedroom, jump on the bed, and lick the face of whichever one of us was "sleeping." Buck was quick to respond even with his limp. Of course, when it was Meg's turn, she scampered in and flung herself, wiggling with joy, on the "sleeping" person. With either dog, Pete and I were sure nobody would ever be allowed to sleep through the alarm clock.

One problem deaf people have is being unable to hear an alarm clock. This dog is being taught to respond to the alarms of three different types of clocks, and to "awaken" the trainer by jumping on the bed and rousing her.

DAVID CUPP

Then we came to that part of the dogs' training where they would have to run back and forth between the source of a sound and the person. One of us would be at the control panel of the bells while the other sat in the next room with Buck on the leash. When the telephone or doorbell was made to ring, Buck was told *"Sound, Buck!"* and was led back and forth to "tell" the other trainer. The cottage is equipped with four different types of doorbells, and Buck had to learn them all. We also taught him to "tell" about the teakettle when it sang and the oven timer when it chimed or buzzed.

We gave the same training to Meg. The senior instructor was very pleased with the progress made by the dogs Pete and I were working with. Truthfully, though, we thought we had two of the best animals in Buck and Meg. Quite a few of the dogs in training flunk out. That doesn't mean they aren't nice dogs—even wonderful dogs; it just means they aren't suitable to be hearing ear dogs. Some dogs, for example, will do just fine in obedience training, where they are told what to do, but don't seem to think independently very well. A hearing ear dog has to be able to identify sounds by itself and act without being told. The animal must be able to say to itself, in effect: "A bell is ringing. That's the smoke alarm. This time I don't just tell my owner about it—I have to lead him out of the house right away." Or, "Come on, owner, get up—the baby is crying. Follow me and I'll lead you to the crib."

One day, about six weeks into the training of Meg and Buck, Mr. Caprano talked to Peter and me about their progress. He asked us many questions about their abilities and temperaments. He was matching up the dogs in the current

87

class with applicants for dogs. One applicant, he told us, was a deaf young woman who was pregnant with her first baby. Her husband was also deaf, and it was hoped that the dog would learn to alert both of them to sounds; but since the wife would be home alone with the baby almost every day, it was especially important that she have a dog.

Mr. Caprano wanted to know if we thought any of the dogs we were working with—Meg, Buck, and a few others—would be suitable for this situation.

Peter and I both answered, "Buck." While Meg was a sweetheart and bright as a penny, and several of the other dogs we knew were real winners, we both felt instinctively that Buck had the intelligence and maturity that would be perfect for this family. When she was asked, the senior trainer agreed.

The way we train dogs to respond to a baby is by means of a doll with a record. The record of a baby crying can be activated from the control panel of the cottage. We placed the doll in the crib and taught Buck to run back and forth between the crib and either Pete or me whenever the "baby" cried. We also created a makeshift playpen on the living-room floor, put the doll in it, and taught Buck to respond to the "baby" there also.

Meanwhile, an owner had also been chosen for bouncy little Meg. Mr. Caprano had selected a twelve-year-old deaf boy named Jeff who needed her not only to increase his independence but to help his parents worry less about him.

The last few weeks of the dogs' three-month training period, Pete and I each took a dog to live in the dorm with us, as did the other student trainers. I had Buck with me. The point was to socialize the animals, to get them used to living with people in houses, instead of in the kennels where they had been living so far.

Some agencies that train dogs for the deaf find it works better to take the trained dogs to the deaf recipients. A trainer accompanies each dog to its new owner and stays for several days helping the dog adjust and learn the sounds of that household. The trainer also teaches the deaf person how to work with his or her dog. This allows the trainer to pick up and solve immediately any problems that might arise between the dog and the new owner in the setting where they will live together.

However, our system of having the deaf owners spend two weeks here seems to work also. From the very first, the dogs live with their new owners in the dorm, and the owners assume all responsibilities for feeding, walking, grooming, and the like. If a problem develops after they go home together, we try to advise them by phone how to solve it; if necessary a trainer who knows the dog will visit them. We keep in touch with all recipients whether they have problems or not.

Sometimes the dogs ignore their new owners for several days here because they still feel allied with the trainers and instructors. Meg and Buck probably thought they belonged to Pete and me. But after the dogs have been fed, walked, brushed, and cared for by their owners, and have been sleeping in their rooms, the bonds of friendship begin to form. We capitalize on this by teaching the dogs to tell their owners instead of us trainers when they hear the sounds in the training cottage.

Occasionally, a dog is returned to us because the situation just didn't work out. A teenage girl had to give back her dog last year. Her mother didn't really like dogs, we suspected, and was not supportive of the dog in the household. The girl was often negligent about taking care of her valuable helpmate, and the mother refused to. Finally, the father insisted that for the dog's sake it should be returned to Collins. We placed the animal very soon with another deaf person.

Finally, the deaf people who were to receive the dogs in Meg and Buck's class arrived. I was in the kennel grooming a dog when the phone rang.

"Vicky, can you please bring Buck to the training cottage?" asked Mr. Caprano. "I'll be there in a few minutes with Mrs. Handelman."

I went to my dormitory room and got Buck, who was sleeping on the rug by my bed. "Come on, fellow," I said as I put his leash on. "You're about to meet somebody important to you." I have to admit I had mixed feelings. I wanted Buck to succeed as a helper and a partner to a deaf person. But I had grown to love him. I realized I had unconsciously been thinking of him as *my* dog. It's very important in this work to develop a mature attitude and not become overly attached to each individual dog.

Buck stood up, stretched himself, yawned, and followed me cheerfully out of the room. I wondered if Mrs. Handelman had been told that her dog had only three legs.

Buck and I reached the training cottage and went in. A very pregnant young woman stood up and walked toward us. When she saw Buck, she stopped. He came up to her in a calm, friendly way and greeted her by slowly and deliberately wagging his tail. None of us said anything, so not knowing what else to do, Buck just sat down.

90

Mrs. Handelman knelt down and looked into his face. Slurp, Buck gave her a big wet kiss. The woman put her arms around him. It looked to Mr. Caprano and me like instant love.

Mrs. Handelman gently touched the stump of Buck's missing leg and asked in sign language what had happened to it. I signed back that I wasn't sure, but Buck might once have been caught in a trap. I told her about how he had been found. She hugged him again and put her cheek against his head. Buck accepted all this with quiet dignity and pleasure.

Mr. Caprano and I felt this was one match that was almost certain to work out well.

Pete was the one who brought Meg to meet her young owner, Jeff. The boy broke into a big smile when he saw Meg. He was so happy he tried to pick Meg up and hug her, but she wiggled out of his arms. She wanted to play. Pete thought she probably sized up Jeff as someone she could have a lot of fun with. But she would learn that she was his working partner as well as his playmate, and do for him the job she had been trained to do.

Sometimes, in the days that followed, Pete let Jeff borrow his bicycle and ride around the grounds of the college, and Meg ran alongside on her leash. I was afraid she would become exhausted trying to keep up with the bike, but she never seemed to and was always ready for more. Nevertheless, Pete made the boy stop frequently and let Meg flop on the grass and rest. He wanted Jeff to learn that the little dog had rights and feelings that must be respected.

One night soon after Jeff's arrival, Pete looked in on him after he had gone to bed. Pete found Meg asleep also—

not on the rug by the bed, but snuggled in the bed with her young owner. Next day, Pete checked this out with Jeff's parents.

"Oh, it's all right," said the mother. "She's a clean little dog, and if she will help my child I'm not going to say no to her sleeping on the bed."

The two weeks that the deaf recipients spent with us were devoted to teaching them all that the dogs knew, and transferring the dogs' loyalty so that they told their owners instead of us when the bells rang and other noises sounded. This went fairly smoothly, with lots of petting and praise when the animals caught on.

Toward the end of her two weeks with us, Mrs. Handelman's husband joined her. He watched her work with Buck with fascination and obvious pleasure. Buck quickly accepted his new co-owner when he realized that this fellow was closely connected with Mrs. Handelman. The deaf couple always signed to each other, and though Mr. Handelman said "Buck," just as she did, most communication with the dog was with hand signals.

One day Mrs. Handelman had an experience with Buck that wasn't in the curriculum. She and her husband had driven into town and taken Buck with them. Mr. Handelman waited in the car as his wife went into several stores to purchase things she needed. Buck was with her. Suddenly, as she started to cross the street, Mr. Handelman saw a car careening around the corner and heading straight toward his wife. She, of course, could not hear the squeal of the tires and roar of the engine, but Buck saw and heard the car coming. In a flash he stiffened and pulled back on the leash, and then as the car descended on them threw his weight against his mistress, knocking her out of the way just in time.

Both Handelmans were still pale when they got back to the college and signed to us what had happened. Buck, of course, looked nonchalant as if he hadn't done anything special. Not bad for a three-legged dog, right?

The last thing we did before the dogs "graduated" and went to their new homes was to fasten on their harnesses—the bright orange harness and leash that identify them as hearing ear dogs. Most dogs for the deaf wear this harness. These dogs are allowed by law in at least thirty-two states to go to all the same public places that guide dogs for the blind can go: restaurants, stores, buses, schools, post offices, and the like. In some places, deaf people with their dogs have run into problems with restaurant managers or storekeepers who aren't aware of the new laws and have refused them admittance. But as the public becomes more familiar with hearing ear dogs, such incidents are bound to decrease.

As Peter and I watched Meg and Buck with their new owners, I felt good but also a little sad. I had become especially fond of these two dogs. I remembered the day Mr. Caprano and I had picked Meg out at the shelter—and how close we had come to overlooking Buck.

I love the work I do with hearing ear dogs, and it makes me feel great to participate in this remarkable training program. I also get a special thrill when I remember that many, if not most, of the dogs selected for this work are homeless and might otherwise lose their lives because nobody adopted them.

The Handelmans waved as they drove off with Buck

sitting up tall on the front seat between them. Buck didn't even look back. He knew he belonged to these two people. He was going home.

"Let's go into town and see a movie," I suggested to Pete.

As we were walking down the driveway toward the gate to the grounds of the college, we saw Mr. Caprano's car pull in. Mr. C. beeped the horn "hello" as he drove past us. Another student was with him, and sitting up on the student's lap, grinning from ear to ear, was a bright-eyed dog. Somewhere there was a deaf person who needed this animal. The dog, like the others in our program, needed a good home and an owner who would love it. Pete and I and the other trainers would help get them together, this year and after we graduated.

7

"PET THERAPY" FOR THE ELDERLY

Mary describes
the program she directs
for providing pets
for the elderly.

One day when I was fourteen years old, my mother asked me to go with her on a mission involving my favorite aunt—actually, my great-aunt, who was eighty-one.

"Please don't make any plans for next Saturday, Mary," my mother said. "I need you to help me move Aunt Martha to the nursing home and get her settled there."

Ever since my Uncle Fred died some years ago, Aunt Martha had been living alone in her little house in a town not many miles from my mother and me. We used to go to see her, and sometimes we brought her to visit us in our apartment in the city. I loved Aunt Martha. She was cheerful and kind, interested in so many things, and always ready to listen.

But she had begun to get frail and forgetful. Once she fell down in her bathroom and couldn't get up. She just had to lie there for hours, frightened and in pain, until her neighbor realized she hadn't seen any sign of my aunt all day and investigated. Luckily, Aunt Martha hadn't hurt herself very seriously or broken any bones. But the incident shook us up.

Another time, my great-aunt put the kettle on to make herself some tea, forgot about it, and almost burned her house down. Sometimes she was confused. "My memory plays tricks on me," she commented. Now, even Aunt Martha herself realized it wasn't safe for her to live alone any more.

My mother had urged her to get someone to live with her so she could stay in her own home, but Aunt Martha didn't care for that idea; and besides, the house was old and continually needed repairs. Mother asked her to come and live with us. She would have found a larger apartment for us all and hired someone to come in and stay with Aunt Martha during the day—my mother worked and I was at school. But the older woman said no, thanks, she didn't like the city. She finally made up her mind that she would prefer to move to Stafford House, a retirement and nursing home where she could furnish her own room and bring many of her favorite possessions. She even knew another elderly lady who lived there. She tried to be optimistic about it, even though it must have been hard for her to make this decision about her life.

Aunt Martha's house was sold, much of her furniture sold or given away, and plans were made to move the things she wanted to bring with her to Stafford House. Suddenly she realized that she wouldn't be allowed to take Pinky.

Pinky was my great-aunt's fifteen-year-old calico cat.

Aunt Martha almost called the whole thing off. "Pinky and I have grown old together," she stated. "She needs me. What will I do without her? I can't part with her now."

Aunt Martha had forgotten, or blocked it out of her mind, that she and my mother had already discussed Pinky early in the arrangements. Pinky was to come and live out her remaining years with us. We already had two cats and a dog, but Pinky was welcome, and Aunt Martha would see her whenever she came to visit us.

When my mother and I got to Aunt Martha's that Saturday, she was sitting in a rocking chair with Pinky on her lap. She turned the cat over to me without a word. She looked so sad.

Then the truck came to get Aunt Martha's stuff, we helped her into our car, and we were off.

Stafford House seemed like a nice place, and the people who worked there were helpful and friendly and tried to make Aunt Martha feel at home. We stayed to help her arrange her furniture and unpack. She kept sending me down to check on Pinky, who was asleep in the cat carrier on the back porch. The rule at Stafford House was "No pets," so we couldn't even bring her inside.

When dinnertime came, Aunt Martha was introduced to a lot of the other people who lived there. It was a shock to me to see all those aging people together. When you're only fourteen, you never think about getting old. Some of these people used walkers or wheelchairs. Some of them seemed befuddled, but I had a conversation with one elderly man, and I remember thinking, This great old guy is as up-to-date as any of the kids I know.

The old lady Aunt Martha already knew seemed happy

97

to see her. Residents could either have their meals in their rooms or eat in the dining room with the others. Mother thought it was a good sign that Aunt Martha chose to have dinner with the others. We went home, promising to visit her soon.

Several times, we brought her to our house, and she was ecstatic to see Pinky. The cat, while she had settled in peacefully enough with us, clearly was glad to see her owner and purred and purred. But Aunt Martha got worse over the summer, and soon the trip became too much for her. Every time we went to see her, she talked about Pinky, though. The nurses told us she sometimes cried and said she missed her cat.

"Do you remember to give Pinky her tablespoon of cream every morning after her breakfast?" she would ask us. Or, "Pinky used to look so pretty curled up on this blue quilt on my bed. . . ." Or, "I hope you never let Pinky out of the house—there's a big dog down the street who chases cats." She forgot we lived in an apartment in the city, not on her old block. I was upset at the change in this wonderful old lady.

Sometimes Mother and I smuggled Pinky in to see her. Aunt Martha was thrilled. She would perk up and seem happier and more alert. But her pleasure in seeing Pinky was always replaced by disappointment when we had to take the cat away again.

I became outraged at the rules. Why should this elderly lady be deprived of the creature who had shared her life for fifteen years, kept her company after Uncle Fred died, and who meant more to her than anyone—even more, perhaps, than my mother and me.

"I think it's mean. I bet a lot of those old men and

women had to give up pets they loved when they moved to that place," I fumed one time on the way home. "I don't see why they can't have animals there. It would be no big deal to arrange it. What's the matter with the people who run Stafford House, anyway?"

"The staff probably thinks pets would be too much trouble to take care of," said my mother. "Or possibly the state health laws prohibit animals in institutions such as nursing homes. Whatever the reason, the rule must certainly break the hearts of many old people."

Aunt Martha went downhill very fast, and in the end Pinky outlived her by a year. The whole experience influenced me strongly. It certainly gave me much insight into the need for the program I'm involved in today, here at the university.

I'm administrative assistant to the dean of the veterinary college, but that's only part of what I do. I'm also the administrator of a program initiated by my boss along with several other professors and a group of students. It's called Animals for the Aged. What we do is take pets on regular visits to the elderly patients in nursing, retirement, and rest homes. We borrow them from the local animal shelter— kittens, dogs, puppies, rabbits, any animals that like to be carried around, handled, and petted. Also, we close off the room where the pets are shown, place them on the floor, and let them romp and play together. They are so comical, wrestling and tumbling about, that the patients just laugh and laugh. Humor is one commodity that's usually in short supply in nursing homes. Laughing so hard gives the patients a lift that lasts for days, we're told.

And—this is the best part—we also place selected animals in nursing homes that have agreed to take them to live there permanently. In the state where I live now, there are no laws prohibiting pets in custodial institutions, so long as the animals don't go into the kitchens or dining areas where food is prepared and served.

Since hundreds of dogs and cats are "put to sleep" every week at our city pound because homes can't be found for them, this project gives some animals their only chance for life. In pounds and shelters across the country, only about 20 percent of the animals waiting for homes are actually adopted. The rest—as many as fifteen million a year—are killed. Yet, there are elderly people who would love to have them.

When I first started to work at the vet school, Animals for the Aged had been organized and a few nursing homes had been visited with great success. Many interested people wanted to join the group, but a central person was needed to pull the activities together and keep track of details. My boss didn't have the time along with his duties as dean of the college.

"Can you take over the organization, Mary?" he asked me. "There's a great need for our program. We know the nursing-home residents enjoy the animals. But we believe that being with these pets does more than just cheer them up and amuse them. Some of us strongly suspect animals are therapeutic to people, that they improve people's mental health and outlook. Groups like Animals for the Aged around the country are conducting studies to document this. Could you be the administrator of our organization?"

"I'd love to!" I told him. Ever since Aunt Martha and Stafford House, I'd wanted to start a project like this. And here, the chance had fallen into my lap!

At the first meeting of Animals for the Aged that I attended, I was fascinated by the people who came, an interesting mixture. There was my boss, a distinguished middle-aged academician who was said to have turned a formerly mediocre veterinary college into a leading one. There were several professors from the veterinary school, including a woman who was said to be an authority on certain feline diseases. There were veterinary and preveterinary students. Then, there were teachers from other departments of the university—a psychiatrist from the medical school, a couple of professors from the school of social work, many students, including a few high schoolers, and several people who worked on campus in secretarial or administrative jobs like mine. What we apparently all shared was a love of animals and a concern for aging people.

We decided to start with all the nursing homes within driving distance of the campus. We divided up the job of approaching the directors to ask if we could bring animals for regular visits. Later on we would work on placing residential pets and on guidelines for allowing new patients to bring their own pets. Eventually we would extend the program to institutions all over the state.

"Dogs and cats! Are you crazy?" was one nursing home director's reaction when we approached him. "They'll cause a commotion and bother people. They'll make messes. They'll bring in diseases!"

Another listened with cautious interest. "Well, I just don't know," she said. "It might cheer up some of our people, and it won't cost us anything. Perhaps we could try a visit and see how it goes."

Still another understood at once what we were trying

to do. "I think you're onto something good," he said. "So many of our elderly residents are listless and withdrawn, no matter how hard we try to get them interested in the projects we have for them and the facilities we provide. Our staff does its best, but we just can't reach some of our old people. I'd be more than willing to try bringing in dogs and cats. Let's start with regular visits and see how that works out. Then we might consider a live-in pet."

I'll never forget my first visit to a nursing home with Animals for the Aged. A veterinary student named Alan and a social-work professor named Mrs. Canter were meeting me at the nursing home. I was going with Carl, who was in charge of humane education at the animal shelter. We were bringing three puppies, four kittens, and two rabbits— and Alan was bringing his own dog, Susie, a beautiful brown mixed breed. Carl drove the shelter's truck with the pets in cages in the back, except for the two rabbits, which were in a box that I held on my lap. We rattled along like Noah's Ark on wheels with the pets barking and mewing in the back. I began to doubt the wisdom of this project—it seemed to upset the animals. But they relaxed and acted perfectly happy when we took them into the nursing home.

This was the first visit of our organization at this particular place, and the nurses and other staff members looked at us with mixed reactions. Several were delighted, some were just curious, and a few seemed disapproving. About sixty elderly men and women were gathered in the lounge, waiting for us. Some were in wheelchairs, others were seated on chairs and couches.

To get things started, Mrs. Canter introduced us, and Alan put Susie through some of her tricks. The audience oh'd and ah'd and clapped in great amusement. Alan also had Susie demonstrate how well she responded to simple obedience commands—sit, lie, stay, heel, and the like. Susie performed like a star. The old people smiled and applauded, even some of the most listless and seemingly more confused ones.

Then the four of us, carrying kittens or puppies or rabbits, started mingling with the audience. I was totally unprepared for the response we got.

"I haven't held anything warm and young and alive in my arms for twenty years," remarked one old lady, cradling a puppy. I told her this pup was still being bottle-fed and asked her if she would mind giving it its bottle, which we had brought. She smiled and said she'd love to. We left the pup with her all evening till we were ready to leave. She enjoyed it so much.

One man seemed fascinated by the little brown rabbit but hesitated to touch it. He stared at it, smiling, for quite a few moments. Then he tentatively reached out and stroked it gently—the wrong way, against the grain of its fur. He was obviously unused to pets, but stimulated enough to respond to it in his way.

"I never yet met a dog that wouldn't come to me," said another frail old gentleman, bending down to stroke one of the dogs. "Come on now, laddie, give me your paw."

"When I was a girl, my father had cats that lived in his bakery," one lady told me, her pale blue eyes lighting up as she played with one of the kittens. "They were so fat and sleek, and I loved to play with them. My father had a horse-drawn bakery truck. The horse's name was Dan, and sometimes my father would give us children rides on Dan's back." She smiled at the memory. "I'd love to be around animals again," she added wistfully.

103

A volunteer has brought a kitten from the animal shelter for a visit to a nursing home, where an elderly woman is enjoying cuddling it.

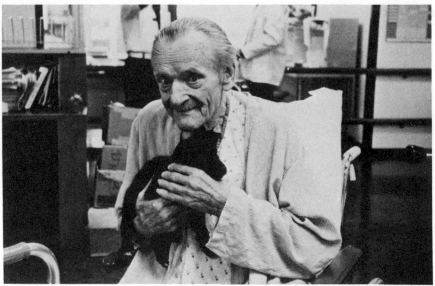

A nursing-home resident is surprised and delighted by a visit from a puppy.

104

One white-haired lady sat silently, expressionless, in a wheelchair off to one side of the room. As I started to approach her, one of the nurses drew me aside. "Don't bother trying to have a conversation with Mrs. Willis," she told me. "She's deaf and doesn't speak. Go ahead and show her an animal, but I doubt if you'll be able to get any response from her at all."

Before I could go to Mrs. Willis, a man plucked at my sleeve and reached up for the kitten I was holding. I knelt beside him for a few minutes while he played with the kitten. When I stood up and turned to go to Mrs. Willis, I was startled by what I saw. There was Carl, leaning over her while she fondled a puppy, having a conversation with her. They were laughing and talking. But Mrs. Willis was supposed to be deaf and noncommunicative! I looked around for the nurse, to see if she saw what I saw, and spotted her with two other staff members, staring at Mrs. Willis and talking to one another in obvious amazement. So much for this elderly lady's silence, I thought. When she found something she considered worth talking about, and someone she decided was worth listening to, she could speak and hear.

"When can you people come back?" asked the nursing home director as we were preparing to leave. "I've never seen these people enjoy anything so much. But more than that, some of them have come out of their shells. Some of them have spoken the first intelligible words we've heard them utter. This is just amazing."

"It would be even more helpful to them if there could be a few pets living on the premises, so they could be with the animals whenever they felt like it," I suggested.

The director said, "Well, we'll see."

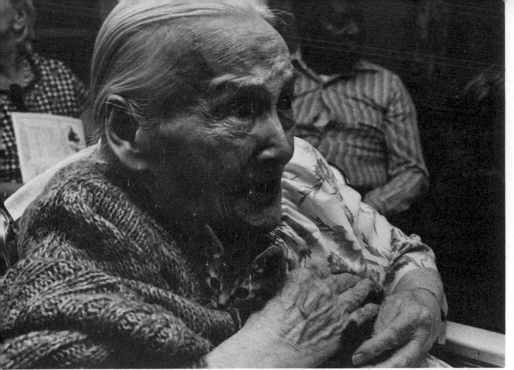

Pets can bring out a warm response in elderly residents of nursing homes. Some withdrawn people speak for the first time when pets are brought to them.

Eventually, this nursing home did take a cat—a sweet, mellow five-year-old neutered tabby that we bailed out of the pound. Carl saw to it that the persons who would be responsible for him knew the basic points of good cat care, and paid several follow-up visits to see how it was working out. Except with a few elderly people who didn't like cats, Felix, as they named him, was a hit. Whenever groups from Animals for the Aged went to visit, Felix was always curled on the lap of one person or another, looking very content.

Even the staff had become converted to the idea of having a resident pet. "Some of these old people have no families, nobody who visits or cares about them," one aide confided to me. "No matter how we try to cheer them up, they are often withdrawn and despondent. But Felix seems to offer so much comfort. He doesn't mind how senile some of the people are. He makes no distinctions whose lap he sits on, so long as they pet him. He offers affection to anyone who will return it, and most do."

Once when we paid a repeat visit to that home, one of the old men took something wrapped in a paper napkin out of his pocket. It was the meat from his plate at dinner.

"I saved it for Susie," he announced. Alan led Susie up to him and told the man to tell Susie to roll over. Susie got into the spirit of things—rolled over, sat up, shook hands, and practically turned somersaults for the little pieces of meat the man gave to her. You'd have thought that dog had never tasted anything so good. The old gentleman said he hadn't had so much fun in years.

We soon placed a little female dog named Phoebe at another home. It came about because of the success of our evening visits. This was a nursing home with very few diversions for the residents except a TV in the small lounge. There were so few patients on hand the first time we went, and the place was so understaffed, I got permission to check some of the rooms to see if anybody needed help to get to the lounge.

In one room, I found one old lady in bed, staring out the dark window. Another patient had told me that this lady almost never got up.

"Why should I bother?" she murmured when I asked her if she would like to be helped to the lounge. "There's nothing to do except watch television. I don't see well enough to read, and I've heard all the stories the other people tell. I know I don't make sense myself sometimes, but some of them are even more regressed than I am. I've outlived myself. I am totally unnecessary."

I asked her if she liked dogs and cats.

"Yes, but there are none here," she sighed. When I told her there was a whole bunch of them in the lounge, she didn't believe me at first. But I persuaded her to get up, come down to the lounge, and see for herself.

When she found herself holding a kitten and petting a dog, this aged lady became a different personality. She talked animatedly, smiled, and had such a good time I was sorry when we had to leave. The next day, I suggested to the director of this place that if they took a pet, I bet that lady would be able to help take care of it. And that's just what happened, to everyone's surprise. This lady would have gone out in all kinds of weather to walk Phoebe, if the staff had let her.

"Phoebe has never caused one problem since she has been here," said the director wonderingly. "The effect she has had on the patients is remarkable. People who never spoke to each other before now gather around the dog and chat. Even the staff has become more caring, perhaps because some of the worst complainers have become less demanding now that they spend so much time playing with the dog. Phoebe goes from room to room in the morning to say hello to everyone. Then she has her walk, and after breakfast she sits with everyone while they read the newspapers or listen to the radio, or in some cases just sit. Phoebe just gives out love all day long to anybody who will accept it— and almost everybody does."

One retirement home has taken a whole collection of

dogs and cats. At this place, the pets don't wander freely about but live in their own quarters on the premises. The old people check them out like books from a library! Every morning, a volunteer takes the dogs on leashes and puts the cats in carriers on a cart and wheels the cart through the halls. Residents wait for them eagerly. They get to pick the animals they want and keep them for two hours. Then an aide comes back and collects those animals and takes them back to their quarters. In the afternoon and evening, they can be checked out for another two hours.

One man likes to turn out the lights in his room and play with a frisky young cat with a flashlight in the dark. This cat will romp around trying to catch the beam of light. Other old people like to hold the animals while they read or watch television or chat with each other. Some just like to have a pet lying around their room.

The dogs and cats have adjusted perfectly to this "work." They spend about six hours a day with the residents. The rest of the time they stay in their own quarters, except for the dogs, which are exercised outdoors twice a day by residents who can get about easily and who sign up for dog walking. Otherwise the animals are taken out by volunteers or staff members who particularly love dogs.

Another place we know of heard about our program and changed their rules so that new patients can bring their own pets. That seems to be working with many benefits and no major problems.

A friend of mine, who is a graduate student, is doing a research paper on the effect of pets in one of our nursing homes. He says quite a bit of research is going on at other universities to document the benefits of pets in custodial institutions such as nursing homes, hospitals, halfway houses, even prisons.

The veterinary school at the university helps us greatly. A faculty member checks over each animal carefully to be sure it is in perfect health before it goes on any visit with Animals for the Aged. And every pet that we place in a home gets any medical care it ever needs. The only real health problem we've encountered is that the residents tend to feed the pets snacks, and the animals get fat.

Not all animals are suitable for nursing-home visits or residence. We are careful to pick only very well socialized, easygoing, calm, friendly pets. The shelter advises us which of the available dogs and cats would be good with elderly people. Then, our veterinarians also examine each animal for its temperament. We have not yet had any bad experiences—cats that scratched or struggled or tried to hide all the time, dogs that became nervous and snapped, nothing like that. As far as I know, no pet has ever been mistreated, either (no resident or staff member would dare!). The worst thing that has happened was when a wheelchair patient accidentally rolled over a cat's tail. The poor cat let out a howl and fled under a bed. It stayed clear of wheelchairs for a few days.

The resident animals that the old people can see every day are far more beneficial than the brief evening visits we are able to make once or twice a month. However, some people among the residents and staff of these homes just do not like animals, period. Their rights must be respected, so the pets are not imposed on them. Even staff members who dislike animals, however, come to appreciate the effect the pets have on most of the elderly people. Their jobs are easier when the residents or patients are happily occupied with the pets.

Several months ago, the wife of one of the professors, Mrs. Jarnigan, came to a meeting of Animals for the Aged with a great idea.

"I bet there are elderly people living alone in their own homes who would love to have pets if they had a little help in caring for them," she said. "Why can't we set up some kind of service for them?"

Mrs. Jarnigan went on to suggest that we could take animals from the shelter, place them with old people, and assign each person a school child or college student volunteer who would be available to help with the animal when needed. For instance, in bad weather when an elderly person can't take his or her dog out, or can't get to the store for pet food or litter, the volunteer could do it.

Carl offered to give the volunteers in the program a course in basic pet care and training before they took on the jobs. Someone else offered to approach stores and supermarkets to ask for donations of pet food for any elderly person who couldn't afford it. My boss said he would guarantee veterinary care for the pets.

We decided to try Mrs. Jarnigan's plan. Her own three youngsters, a girl of sixteen and two boys aged thirteen and eleven, signed up right away. When young Paul Jarnigan's Boy Scout troop heard about it, some of those children joined. Their Scout leader promised them points for this. The principal of the junior high asked for volunteers and collected a few more kids, especially when he offered some kind of extracurricular credit for it. Several high-school and college students volunteered just because they liked the idea of helping. About thirty young people showed up for Carl's first pet care and training class.

Meanwhile, we spread the word about our project through senior citizen centers, church groups, social workers, anybody who had anything to do with the aged. Some twenty elderly men and women were eager to join, including four who had pets of their own. These four had been on the verge of having to give up their pets because it was becoming impossible to take proper care of them.

"I was thinking I would have to have poor Buster put to sleep," one old gentleman, Mr. Florio, told me. "He's too old and crotchety to be put up for adoption. Yet, I can't take him out when the weather is bad. Last year I slipped on the ice, broke my hip, and spent three months in the hospital. I had a good neighbor family who took care of Buster when I couldn't, but they moved away. I didn't know what else to do about him. Now, I can keep him. We need each other. It's just us two old fellows alone together, right, Buster?" He fondled his old dog's head, and the dog licked his hand.

Paul, who was eleven, stopped in after school the next day to meet Mr. Florio and Buster. Unfortunately, Buster blew it. The dog would have none of Paul, growled and grumbled and snapped when Paul tried to take him out for a walk.

"I wouldn't mind walking Mr. Florio, he's so nice, but I don't want to have anything to do with Buster—I'm scared of him," Paul told his mother.

"Let me try," said David, Paul's older brother. "Maybe I can get Buster to cooperate."

Mr. Florio was almost ninety years old, but except for being rather deaf, he was in complete command of his senses and had a playful wit. David got a kick out of the stories the elderly man told him. With a lot of patience and some coaching from Carl, David took on Mr. Florio's grouchy little dog. He stopped by every few days to chat with the old man and let the dog check him out. After several weeks, Buster suddenly went to David, licked his hand, and lay down with his head on David's boot. David knew then he had won him over. Mr. Florio was so happy he almost wept. Now Buster trots out the door with David without hesitation.

The dogs and cats that were taken from the shelter and placed with the old people who wanted them were carefully chosen and worked out well—all except in the case of one lady who hadn't told us she had asthma. The hair from the cat we gave her irritated her condition, and we had to take the cat back. However, a pet store gave us a parakeet for her, and she loves it.

Recently, one of the members of Animals for the Aged contacted a children's hospital and got permission to take pets there. We went to a downstairs waiting room with a bunch of kittens and puppies—and Susie, of course. All the ambulatory and wheelchair children gathered around in great excitement. One pale little girl with her back in a brace didn't speak at all, but she snuggled a kitten. A nurse told us it was the first time she'd seen the child smile.

The good that animals do for people in institutions seems endless. We'll have to change the name of our organization—we are reaching more than the aged. I love the work we do, and I know it means a great deal to many people. There's a rising interest and much research on the beneficial effects of pets. My boss plans to deliver a paper on our program at an international meeting in London.

In states where the health laws stand in the way of allowing pets in nursing or rest homes, the laws must be changed. The notion that animals are automatically "unsanitary" or "unhealthful" is not based on fact. Many elderly people would not only enjoy having pets but would benefit greatly from having them.

We have heard of other groups around the country with programs similar to ours, and next year we plan to hold a symposium and compare notes. We know we're onto something wonderful.

FURTHER
INFORMATION

While there are many good books on guide dogs for the blind, not much has been published about other fields of work that involve training or using animals to help handicapped people. Much of this work is new. However, an enterprising person can find magazine articles on these subjects in *The Reader's Guide* or in the vertical file index in libraries, perhaps listed under "companion animals" or "service animals," "pets," "guide dogs," "handicapped and animals," and "aged and animals."

To enter some of these fields, you need two to four years of college; others do not require college. Familiarity with animals is essential. You usually do not need experience with handicapped people—you'll get that during your training. High schoolers would do well to take courses in the biological sciences and get as much experiece with animals as possible. 4-H activities also offer good preparation.

Organizations that train animals for the handicapped often have informational material available on request. It would be impossible to list all of these agencies here, but the following should give an interested person a start.

General

Films for Humane Education, a useful book edited by Ronald Scott and Jean Stewart, provides complete information about over one hundred films on animals, including service and companion animals. A valuable aid to librarians, teachers, and group leaders of many kinds, this handy loose-leaf manual offers clear details about each film: type, format, description, possible discussion subjects, and sources for borrowing, renting, or buying.

The book can be ordered for $4.75 plus $1 postage and handling charge, from Argus Archives, 228 East 49th Street, New York, New York 10017.

People-Pet Partnership Program, a monograph by Linda Hines, tells how to establish a good community-based, volunteer-staffed activity that involves animals in therapy for the handicapped, aged, and lonely and conducts projects that communicate the value of animals to people. Tells how to set up a therapeutic horseback-riding program, teach humane education in schools, visit nursing homes with pets, and other projects, based on the experiences of the successful People-Pet Partnership Program at Washington State University (Pullman, Washington). The publication is available for $5 from the Latham Foundation, Latham Plaza Building, Clement & Schiller, Alameda, California 94501.

Pet Facilitated Therapy: A Selected, Annotated Bibliography, lists books, magazine articles, and other published material about PFT activities. This booklet is a useful tool for persons who are interested in research or who are trying to establish a pet-facilitated therapy program. Can be ordered for 50¢ from the American Humane Association, 9725 East Hampden Avenue, Denver, Colorado 80231.

Aide Dogs

Some professional dog trainers will train dogs to serve handicapped people who are confined to wheelchairs. Some will teach handicapped persons how to train their own dogs in obedience and possibly to perform some helping tasks. At the time of this writing, most of this work is still taught by the apprentice system, and the way for a young person to learn it is from an established trainer. A broad general education plus experience with dogs is excellent preparation, but as Katie says in Chapter 1, you also need energy, mature judgment, ability to innovate—and lots of patience.

Persons interested in aide dogs can contact the following:

Aid Dogs for the Handicapped Foundation, Starr R. Hayes, director. Dogs custom-trained to serve disabled individuals. 1312 Bergan Road, Oreland, Pennsylvania 19075. (215) 233-2722

Handi-Dogs, Alamo Reaves, director. Special classes in which handicapped people (including deaf) are taught to train their own dogs. P.O. Box 12563, Tucson, Arizona 85732. (602) 326-3412 or (602) 325-6466

Able Dogs, Sue Myles, director. Special classes in which handicapped people (including deaf and mentally retarded) are taught to train their own dogs in obedience. 502 31st Street, Newport Beach, California 92663. (714) 673-3085

Therapeutic Horseback Riding

A young person wishing to go into this work would find that, in addition to riding experience and familiarity with horses, college courses in equestrian science or veterinary technology would be extremely valuable. However, trainees from many backgrounds—teaching, for example, or physical therapy—may be accepted, depending on the individual.

There are approximately two hundred centers in the United States and Canada that offer therapeutic riding. Some also have instructor training programs. Information about them can be obtained from:

North American Riding for the Handicapped Association
Leonard Warner, executive director
P.O. Box 100
Ashburn, Virginia 22011
(703) 471-1621 or (703) 777-3540

A large, illustrated paperback book, *It Is Ability That Counts*, by Lida L. McCowan, is a thorough training manual on therapeutic riding. Available for $6 from the Cheff Center for the Handicapped, R.R. 1, Box 171, Augusta, Michigan 49012.

Ability, Not Disability, a 23-minute documentary color film on therapeutic horseback riding, is available in 16 mm or ¾" video cassette. Filmed at the Cheff Center for the

Handicapped, the film offers vivid evidence of the benefits of this recreational therapy. $10 per week rental charge. Can be ordered from: The Latham Foundation, Latham Plaza Building, Clement & Schiller, Alameda, California 94501. (415) 521-0920

Guide Dogs for the Blind

Like other work training dogs for handicapped people, guide dog training requires prior experience with dogs, and most agencies prefer at least two years of college. Veterinary technician training is considered good, as is a background of kennel work or in 4-H or the K-9 Corps. Maturity, verbal skills, and physical fitness are especially important.

Guide dog schools that train dogs for the blind are listed below. For information regarding apprentice programs for trainers, contact the schools directly.

Fidelco Guide Dog Foundation
P.O. Box 142
Bloomfield, Connecticut 06002
(203) 243-5200
Guide Dog Foundation
371 Jericho Turnpike
Smithtown, New York 11787
(516) 265-2121
A 16-mm sound film,
Out of the Shadows,
can be borrowed free of
charge from this foundation.

119

Guide Dogs for the Blind
P.O. Box 1200
San Rafael, California 94902
(415) 479-4000

Guiding Eyes for the Blind
Granite Springs Road
Yorktown Heights, New York 10598
(914) 245-4024

International Guiding Eyes
5528 Cahuenga Boulevard
North Hollywood, California 91601
(213) 877-3937

Leader Dogs for the Blind
1039 South Rochester Road
Rochester, Michigan 48063
(313) 651-9011

Pilot Dogs
625 West Town Street
Columbus, Ohio 43215
(614) 221-6367

The Seeing Eye
P.O. Box 375
Morristown, New Jersey 07960
(201) 539-4425

Greff: The Story of a Guide Dog, a book by Patricia Curtis, tells children about the early life and rigorous training of a young dog who becomes a guide and companion to a blind person. Published by Lodestar Books, 2 Park Avenue, New York, New York 10016; 1982; price $9.95.

this work. Specific information should be obtained by writing to the organizations themselves.

American Humane Association, 1500 West Tufts Avenue, Englewood, Colorado 80110. (303) 762-0342 (voice); (303) 789-1278 (TTY)

Dogs for the Deaf, Applegate Behavior Station, 13260 Highway 238, Jacksonville, Oregon 97530. (503) 899-7177 or (503) 899-7542

Guide Dog Foundation, 371 Jericho Turnpike, Smithtown, New York 11787. (516) 265-2121

International Hearing Dog, Inc. (Agnes McGrath, director), 5901 East 89th Avenue, Henderson, Colorado 80640. (303) 287-3277 (voice/TTY)

New England Education Center, Hearing Ear Dog Program, Bryant Hill Farm, 76 Bryant Road, Jefferson, Massachusetts 01522. (617) 829-9745 (voice/TTY)

San Francisco SPCA, Hearing Dog Program, 2500 16th Street, San Francisco, California 94103. (415) 621-2174 (voice/TTY)

Cindy, a Hearing Ear Dog, a book for children by Patricia Curtis, with photographs by David Cupp, tells the story of a dog who is trained to be a hearing ear dog and valued companion for a deaf teenager. Published by E. P. Dutton, 2 Park Avenue, New York, New York 10016; 1981; price $10.25.

"Pet Therapy" for the Elderly

Many humane organizations, such as the ASPCA in New York City, the AntiCruelty Society in Chicago, and SPCAs,

Humane Societies, and Animal Rescue Leagues in other cities, have programs for taking pets on visits to nursing homes and hospitals or for placing resident pets in such institutions. Most of these organizations welcome volunteers. Young persons interested in this rewarding activity should inquire at their local humane organizations to find out if there are "pet therapy" programs and what the requirements (age, for example) are for volunteers.

An invaluable monograph for anyone interested in the field is *"Pet Therapy": A Study of the Use of Companion Animals in Selected Therapies*, by Phil Arkow. Describes "pet therapy" and its benefits, tells how to establish and implement a pet-therapy program, including how to avoid or deal with its problems. Available for $3.50 from the Humane Society of the Pikes Peak Region, P.O. Box 187, Colorado Springs, Colorado 80901.

Professionals already involved in nursing-home work would find this booklet helpful: *Animals in the Nursing Home: A Guide for Activity Directors*, by Cappy McLeod. Can be ordered directly from Ms. McLeod, 16 West Willamette #7, Colorado Springs, Colorado 80903; price $6.50.

A comprehensive book on the use of animals in therapy for the elderly is *Animals, Aging, and the Aged*, by Leo K. Bustad, D.V.M. An overview of the contributions of companion animals to the mental health of older people, the book includes a discussion of the problems of the aged, plus practical information on selecting pets, guidelines for placing pets, and the like. Published by the University of Minnesota Press, 2037 University Avenue S.E., Minneapolis, Minnesota 55414; 1981; price $19.50.

INDEX

Italic page numbers refer to captions.

Able Dogs, 117–118
Aid Dogs for the Handicapped
  Foundation, 117
aide dogs, 1–14
  agencies for, 117–118
  breeds used as, 10
  children's self-esteem and,
    2–3, 6–7, 8
  objects retrieved by, 9, *9*,
    11–12
  as physical supports for handi-
    capped people, 10
  temperament needed for,
    13–14
  training of, 2, 6–7, 8, 10–12
American Humane Association,
  85, 123

American Society for the Pre-
  vention of Cruelty to Animals, 123
Animal Rescue Leagues, 124
animals:
  adoption rate of, 100
  changing views on, xiii
  children's feelings and, *54*,
    58–60, *61*
  handicapped people's speech
    and, 13
  noncritical acceptance of chil-
    dren by, 55, 69, 73
  patient-therapist communica-
    tion aided by, 63, 69–71
  professional opportunities
    with, xiii
  rights of, 53, 60, 62
  utilization vs. exploitation of,
    xi–xiv

animals, *continued*
    volunteer work with, xiii–xiv,
      111–113
    see also pets
Animals for the Aged, 99,
    100–114
    goals of, 101
AntiCruelty Society, 123
Apple Tree Farm, 52–60, *54*, *61*,
    62, 64–66
    background of children on, 52,
      53
    care of farm animals on, 53
    founding of, 52
    goals of, 52, 64
    riding classes at, 55–57
    rights of animals taught at,
      60, 62
    as working farm, 53, 57–58
    see also farm animals, Apple
      Tree
autistic children, therapeutic
    equitation and, 26

blind people:
    as applicants for guide dogs,
      44–45
    guide dog training and, 34–36,
      37–38, *45*, *47*, 48–49
    matching guide dogs to, 44
    offering help to, 49
blood pressure:
    of dogs, 72
    of humans, 72–73

Cheff Center for the Handi-
    capped, 24–26
    intensive training program of,
      25

Collins Junior College, hearing
    dog program of, 80, 81
Corson, Elizabeth and Samuel,
    74

Davies, John A.; 24
deaf people:
    hearing ear dogs for, 77–94
    population of, 80
    problems of, 80–81
    sign language vs. lip reading
      for, 81
directed retrieve, 11–12
dogs:
    children's self-esteem and,
      2–3, 6–7, 8
    crutches retrieved by, 5, *5*
    effect of stroking on, 72
    hunting, 11–12
    objects retrieved by, 9, *9*,
      11–12
    as physical supports for handi-
      capped people, 10
    with retarded people, 73–74
    see also aide dogs; guide dogs;
      hearing ear dogs
Dogs for the Deaf, 123
dog trainers:
    for aide dogs, 1–14, *5*, *9*
    for guide dogs, 36, 38–40, *45*,
      47

elderly, *see* "pet therapy" for
    the elderly
empathy, *61*, 62

farm animals, Apple Tree,
    51–66

126

farm animals, *continued*
  attachment of children to,
    58–60
  body language and, 53
  learning respect for, 53, *61*
  learning self-control with,
    55–57
  nonverbal children and, 53,
    54
  as playmates, 53–55
  pregnancies of, 58–60, 65–66
  prizes won with, 55
  slaughtering of, 57–58
Fidelco Guide Dog Foundation,
  119
4-H competitions, 55
frogs, rights of, 60, 62

German shepherds, 42
Gillespie, Ward, 3–6
golden retrievers, 10, 42
Green Haven, 26–33
  children's riding program at,
    29–30
  horse training at, 30–32
Guide Dog Foundation, 119, 123
guide dogs, xiv, 34–50, *45, 47*
  agencies for, 46
  breeding of, 42–43
  disposition of, 42
  foster homes for, 42
  matching of, 44
  new homes and, 46
  obeying vs. disobeying learned
    by, 43
  petting as distraction to,
    48–49
  requirements for trainers of,
    38–40

guide dogs, *continued*
  screening of applicants for,
    44–46
  on subways, 48
  training of, 40–41, 43
Guide Dogs for the Blind, 120
guide dog trainers, *45, 47*
  education of, 36, 41–42
  requirements for, 38–40
Guiding Eyes for the Blind, 120

handicapist attitudes, 49–50
handicapped people:
  animals trained for, xii–xiii,
    xiv
  rights of, xiv, 2
  society's view of, 2, 49–50
Handi-Dogs, 117
hand signals, 2, 6, 40, 82–83
harnesses, for hearing ear dogs,
  93
Hartel, Liz, 24
healing process, pets and, 67–76
  research on, 71–74
health, pets and, 71–72, 74, 76,
  100
health laws, need for changes in,
  113–114
hearing ear dogs, xiv, 77–94
  agencies for, 85, 122–123
  alarm clocks and, 83, 85, 86,
    86
  baby's crying and, 83, 87, 88
  doorbells and, 83, 85, 87
  hand signals taught to, 82–83
  harnesses for, 93
  household sounds and, 83–85,
    84, 87
  intelligence of, 87

127

hearing ear dogs, *continued*
  new laws and, 93
  new owners and, 89–93
  outdoor sounds and, 85
  smoke alarms and, 83, 85
  telephone and, 83, *84*, 87
  training system used for, 83
heart disease, pets and, 71
horses, horseback riding:
  building confidence with,
    55–57
  see also therapeutic equitation
hospitals, children's, xiv, 113
"human / companion animal
    bond," xiii, 71–72, 76
Humane Societies, 124
humor, in nursing homes, 99

International Guiding Eyes, 120
International Hearing Dog, Inc.,
    123

Labrador retrievers, 42
Leader Dogs for the Blind, 120
Lee, David, 63
lip reading, 81

McCowan, Lida, 25
McGrath, Agnes, 85, 122

New England Education Center,
    123
North American Riding for the
    Handicapped Association,
    25, 118
nursery schools, use of pets in,
    68–69, 74–75
nursing homes:
  health laws and, 113–114

nursing homes, *continued*
  need for humor in, 99
  use of pets in, xiv, 98–110

Oaklands Riding Center, 15–24

"Pet-Facilitated Therapy,"
    67–76
  for preschoolers with emo-
    tional problems, 67–71,
    74–76
  research and, 71–73, 74
pets:
  in a children's hospital, 113
  for the elderly, 95–114, *104*,
    106
  with elderly living alone,
    111–113
  in nursery schools, 68–69,
    74–75
  in nursing homes, 98–110
  physical health improved by,
    71–72, 74
  in psychiatric hospital/prison,
    63–64
  resident, 106–111
  talking to, 72–73
  see also animals
"pet therapy" for the elderly,
    95–114, *104*, *106*
  Animals for the Aged and, 99,
    100–114
  animal temperament in, 110
  children as aides in, 111–113
  people living alone and,
    111–113
  personality changes in, 103,
    105, 108

128

"pet therapy" program, in psychiatric hospital / prison, 63–64
Pilot Dogs, 120
Pony Riding for the Disabled, 24
punishment, in animal training, xii, 40

research on "pet therapy," xiii, 71–74, 76, 109–110
retarded people:
  dogs and, 73–74
  horses and, 25–30, *28*, 32–33
riding classes, 55–56

San Francisco SPCA, 123
Seeing Eye, The, 120
Seeing Eye Dogs, *see* guide dogs
self-esteem:
  dogs and, 2–3, 6–7, 8
  horses and, 22, 25, 33, 57
sign language, 81–82
Skeezer, 69, *70*
Snow, Alexandra, 16–33

subways, guide dogs on, 48

therapeutic equitation, xiii–xiv, 15–33, *17, 18, 28*
  as ancient medicine, 24
  emotionally disturbed children in, 22–24, 25–26
  grooming in, 16–18, 22
  physically handicapped people in, 16–22, 25
  retarded adults in, 25–26
  retarded children in, 26–30, *28*, 32–33
  self-esteem gained in, 19, 22–27, 55–57
  strength developed in, 19, 21, 29–30
  teaching children in, 22–24, 29–30
  training horses for, 30–32
turtles, rights of, 60, 62

vegetarians, 57
volunteers, xiii–xiv, 111–113
  see also specific agencies

ABOUT THE AUTHOR

PATRICIA CURTIS, a former editor at *Family Circle* magazine, is the author of *Greff: The Story of a Guide Dog; Cindy, A Hearing Ear Dog;* and three other books about animals. "I believe we are experiencing a slowly rising awareness of the bonds between human beings and companion animals," she says. "And I hope this appreciation will improve the ways people treat animals." Ms. Curtis lives in New York City.